I0810097

IMAGES
of America

TUCSON RADIO

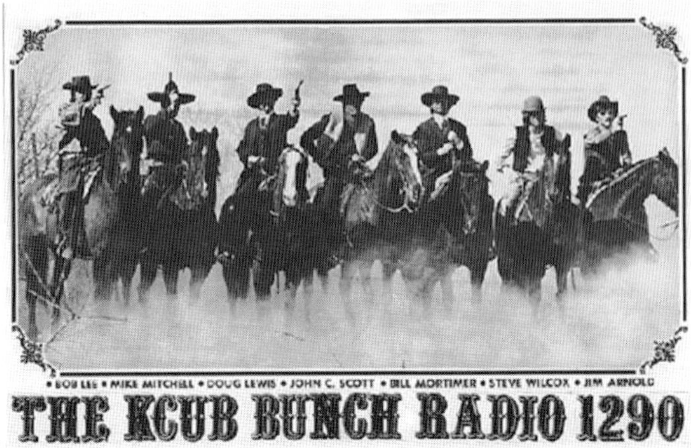

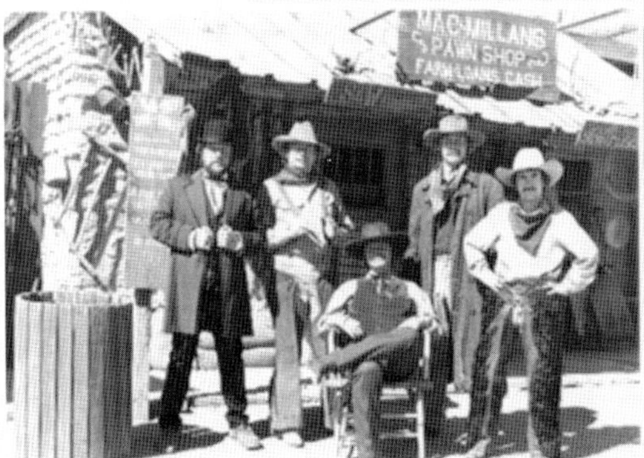

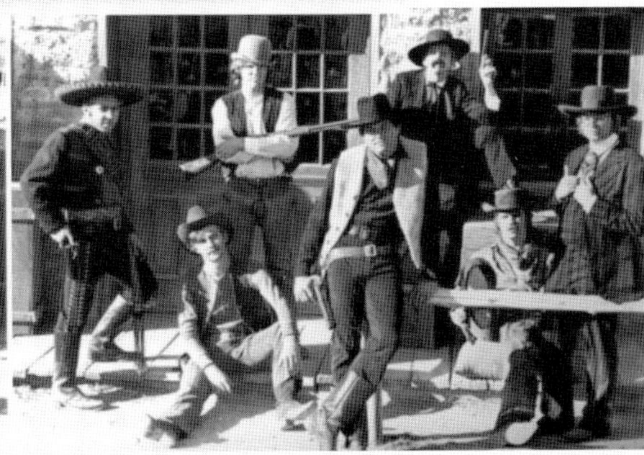

At the top in these three images of Tucson, radio, and the Old West is the on-air staff of KCUB radio, posing on horses at the Old Tucson movie set in 1972. Two of the "KCUB Bunch" riders went on to be inducted into the Arizona Broadcasters Hall of Fame: Bob Lee, on the far left, and Jim Arnold, on the far right. In addition, KCUB owner Jim Slone is an inductee. John C. Scott continued as a newscaster and radio talk-show host, and Bill Mortimer, a Tucson native, retired after working at several local stations. The dismounted "bunch" in the lower left picture taken at Old Tucson in 1978 are, from left to right, Jim West, Bill Mortimer, Doug Brannon, Sunny Jim Arnold, and Dave Slavin. The "bunch" on the right are the same ones in the top picture, but dismounted and looking "desperate," from left to right are (seated) Sunny Jim Arnold and Bob Lee; (standing) Mike Mitchell, Steve Wilcox, John C. Scott, Doug Lewis, and Bill Mortimer. The pictures were produced by Norma Slone. (Jim Slone.)

On the Cover: Frank Kalil was a "young, wide-eyed curly headed kid staring at me through the window at KTKT," according to radio legend and 1950s KTKT personality Chuck Blore. Before long, Kalil was the number one radio personality in Tucson, with his show of Top 40 music, comedy, and promotion events airing from 3:00 p.m. to 6:00 p.m. He became a station owner at KAIR, then a businessman, growing his company into the top brokerage firm for radio-station ownership in the country, headquartered in Tucson. (Chuck Simms.)

IMAGES
of America

TUCSON RADIO

Russ Jackson
and C.J. Brown

ARCADIA
PUBLISHING

ISBN 978-1-4671-3140-7

Published by Arcadia Publishing
Charleston, South Carolina

Printed in the United States of America

Library of Congress Control Number: 2013948138

For all general information, please contact Arcadia Publishing:
Telephone 843-853-2070
Fax 843-853-0044
E-mail sales@arcadiapublishing.com
For customer service and orders:
Toll-Free 1-888-313-2665

Visit us on the Internet at www.arcadiapublishing.com

To Susan and Gerry, who continue to support our endeavor to know more about Tucson radio and its people.
To all the veteran Tucson broadcasters who made the sound that inspired listeners for so long and continue to keep radio "alive" in the Old Pueblo, and to all the radio listeners who keep listening.

Contents

FOREWORD

The broadcasters who people remember best are those hometown voices that helped them navigate the changes in life, through school and into their careers. In effect, those voices make every market unique. Tucson is no exception. A mid-sized market deep in Southern Arizona, where radio first arrived in March 1922, Tucson radio has developed its own distinct desert flavor. In fact, when one station offered a format, talent, and sound like that of a top Los Angeles station, it was unable to top the market. Over the years, Tucson has seen all the typical radio promotions and remote broadcasts. Nevertheless, few markets can boast broadcasts of local camel races, or a station owner sitting at a card table in front of his station, watching commuters drive to work—and talking about it, live, on the air.

With more stations per capita than most markets, Tucson supported a vibrant radio dial, developing a number of personalities who gained national stature, including Chuck Blore, Jefferson K (aka Shadoe Stevens), Frank Kalil, Lou Waters, and Terry MacDonald. Unfortunately, as in many cities, major consolidators now preside over complexes of empty studios, featuring mostly automated voice tracking and satellite programming from somewhere else. Still, a few stations strive to stay local and live.

So, read on. Dial around and, if you listen carefully, you can still hear the echoes and memories of Tucson radio.

—Barry Mishkind, the eclectic engineer

ACKNOWLEDGMENTS

The authors thank all the Tucson broadcasters who graciously provided materials, photographs, and feedback that helped us put together this illustrated history of a medium of sound. Special thanks go to veteran Tucson broadcaster Larry Schnebly, whose industry connections played a vital role; 50-year Tucson radio broadcaster and station owner Jim Slone, who shared his career in photographs and promotion pieces; Frank Kalil for his hospitality, time, and materials; "Sunny Jim" Arnold, whose career also went from disc jockey to station manager, shared his time, career, and staff in a memorable visit to the Journal Broadcast Center; Barry Mishkind (the eclectic engineer) and his great website, www.oldradio.com; Ray Lindstrom, for his material and website, www.ktkt99.com; Bill Mortimer, Bob Lee, Chris Borden, Shadoe Stevens, John Wasley, Roger Galloway, Jerry Stowe, Frank Casanova, and Tom Lang; Mrs. Gene Autry, for the great photograph of her late husband; Laura Hoff at the Arizona Historical Society in Tucson; MaryLee Carter, for the photographs and memories of her father, George Bradley; Ron Weiss, for calling our attention to that magnificent KCNA clock he proudly displays in his store and for providing the photograph of his mother-in-law, Priscilla Sanders; Shay Oliver for sharing photos; Bob Orozco, whose father, Jacinto Orozco, was a pioneer in Spanish-language radio, for opening the family archives; Joe Howell, at Tierra Right of Way Services Ltd.; Buzz Jackson of KIIM, for sharing his huge photograph collection; Lynn Spasoff and Herb Crowe of Cumulus Broadcasting, for the illustrations; Clear Channel Broadcasting's Chris Abbott, for hosting a tour of its facilities, and Nikki Van Doran, for the contacts with Jeff Davis and Randy "R Dub!" Williams; John Kelley of Arizona Public Media; Hector Youtsey, for photographs of the unique radio station; UA Wildcat sportscaster Brian Jeffries; Doug Martin of Good News Broadcasting, for the tour of his stations and for the materials about them; Pearl Ford-Fyffe of the Ad Federation, for the photograph of Phyllis Ehlinger; a special thanks from author C.J. Brown to Chuck Simms and Jim Chrissman, both of Tucson, for everything they have shared with me; and Larry Schnebly, again, for hosting both of us at the "Coffee Crew."

We also thank our editor, Alyssa Jones, for her patience and expertise in helping us sort out the technicalities of book publishing and photograph selection. Both of us thank each other for having the foresight to collect so much memorabilia over the years in the form of photographs and images, which made doing this book more interesting every day. C.J. Brown saved the KVOA 1940s photograph set and the KCNA 1949 promotion booklet. Russ Jackson's mother, Pauline Jackson, saved the 1946 KVOA NEWS image and the 1959 KTKT Top 40 survey. Items marked (CJB) are from the C.J. Brown collection, and those marked (RJ) are from Russ Jackson's collection.

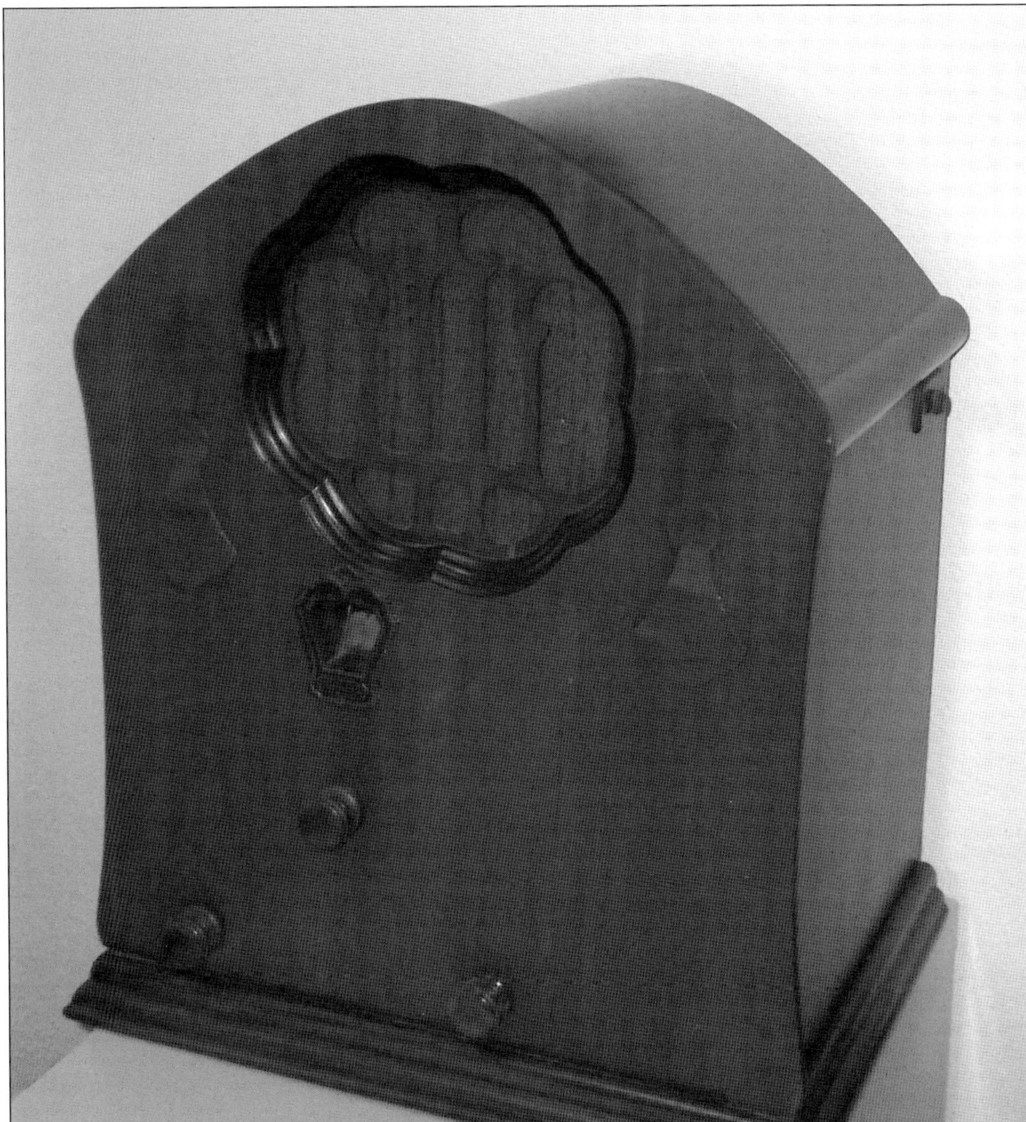

Radio as a medium has been in existence since 1920. The first official broadcast is thought to have been the 1920 presidential election results, aired on KDKA in Pittsburgh, although experimental stations and WWJ Detroit were on the air before that. Listeners began to gather around their battery-powered crystal sets with earphones, until radios like the Clarion Jr. seen here were available for home use in the late 1920s and 1930s. Listeners tuned in to hear programs, some 15 minutes long, some an hour or longer. Each program was an entity unto itself, produced live, with the audience hearing the performance, the event, or the speech as it was happening. The world was being heard in homes for the first time. Radio technology has come a long way since its infancy, as transistors replaced tubes, portability became available to all, and the Internet and social media came along. Today, Tucson radio listeners can choose from over 50 stations on AM and FM. Radios came to the automobile, as a luxury, in the 1930s; now, they are standard in all vehicles, offering listeners their choice of entertainment, news and sports, and traffic reports. While the industry has progressed dramatically over the decades, one thing remains the same: radio serves as a companion. (RJ.)

One

In the Beginning

Two Radio Stations

The history of Tucson radio started with two stations, KTUC and KVOA. They had Pima County all to themselves until 1947. As Barry Mishkind writes on www.oldradio.com, "The oldest surviving local station, KTUC, is a descendant of the *Tucson Citizen*–owned station KGAR. Another early local station was KVOA, now KCUB, which was licensed on May 27, 1929. KVOA lives on as the call sign for television's channel 4." KGAR became KTUC and was eventually assigned to 1400 on the dial. It was sold to KOY in Phoenix, and it carried CBS network programs as well as KOY-produced programs and newscasts that were sent to Tucson on the Arizona Network.

KVOA was assigned to 1290 and was purchased by the Phoenix broadcaster KTAR in 1939. KVOA was now an affiliate of NBC and carried KTAR-produced programs and newscasts that were sent out on the Arizona Broadcasting System network of nine stations around the state. Both Tucson stations produced local programs that were scheduled around the network offerings. During World War II radio came of age, becoming an important source for news. Morale-building programs were heard nationwide, and both KTUC and KVOA endorsed special local programs that urged listeners to buy war bonds. This chapter spotlights the many programs and personalities that defined an era and examines the substantial growth that took place following the war, as the Old Pueblo went from a small town to a city of 50,000 residents. Radio grew along with it.

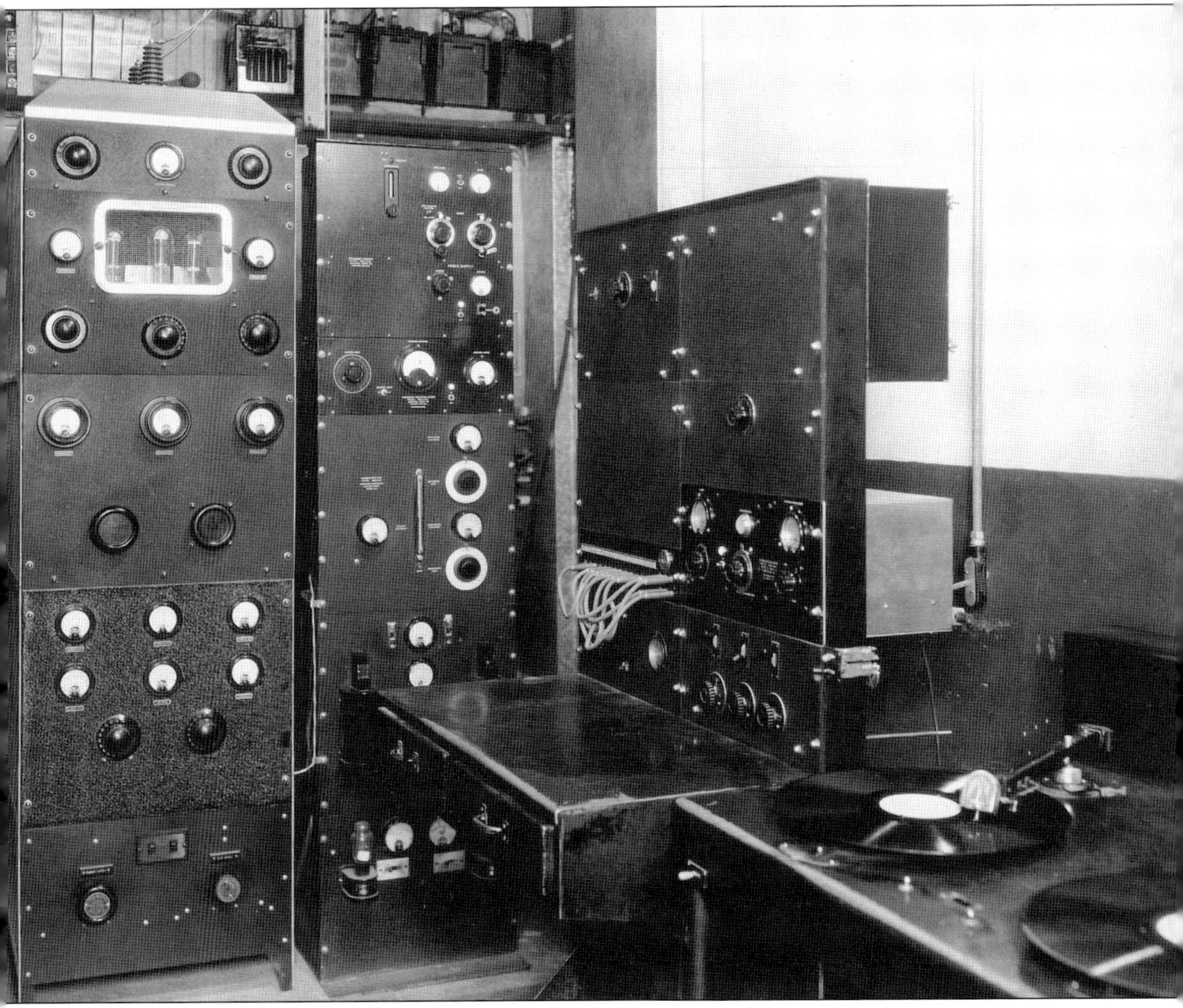

After its purchase from the *Citizen*, radio station KGAR used this equipment in 1935 in its owner's building, at 142 South Sixth Avenue, the Tucson Motor Service Company. A fire station was next door. Tucson native Gene Stough remembers that KGAR had a plate-glass window on the east side of the studio. As a boy, he watched announcer Victor Gelard sitting with a microphone on the desk in front of him. Stough remembers thinking, "Why could I hear him at home and not be able to hear him while watching?" That curiosity led to Stough's career as a broadcast engineer, working in Tucson from 1950 to 1976 at KCNA, and then in California. (Arizona Historical Society, Buehman Photograph Collection.)

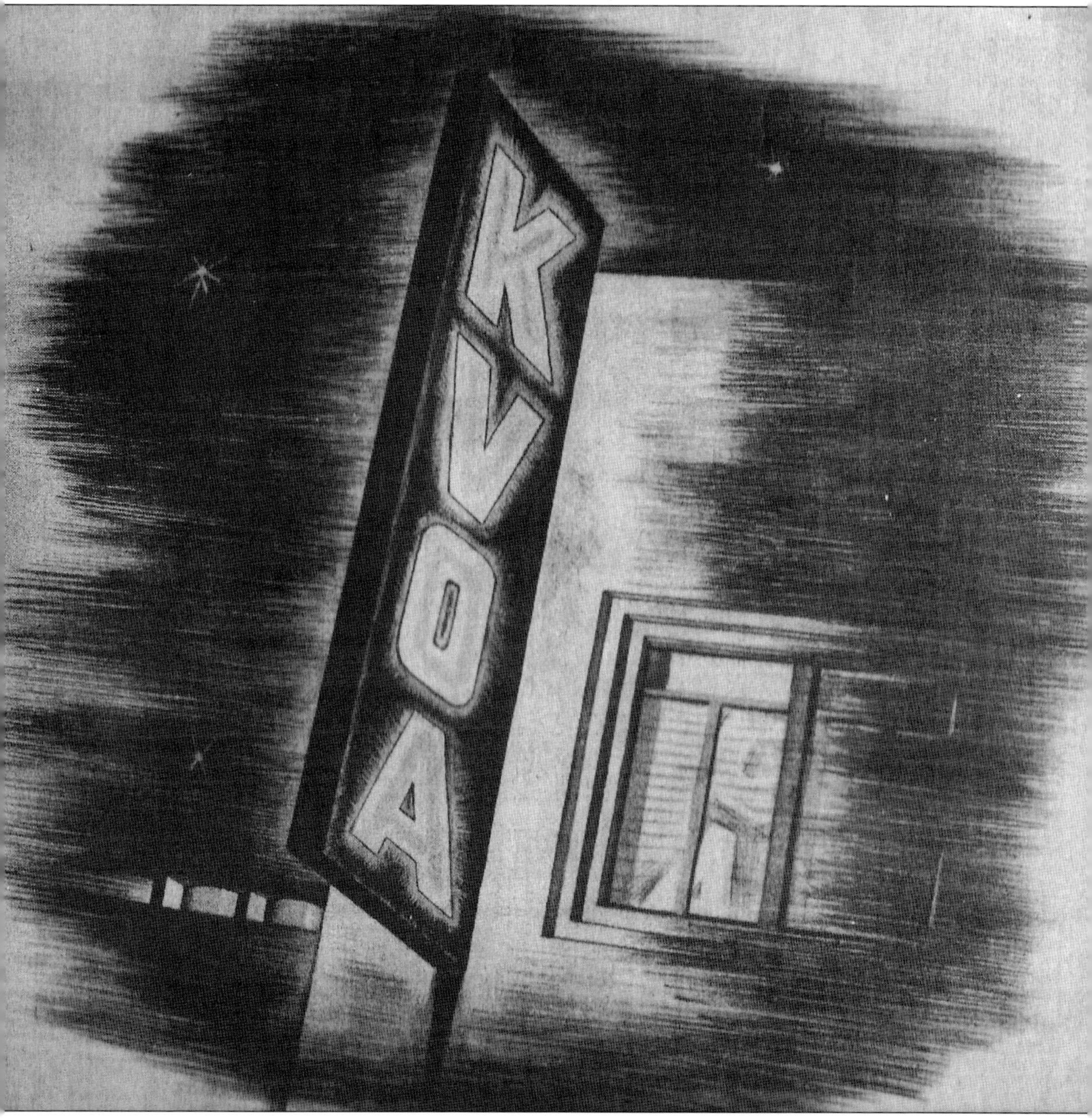

In 1940, KVOA moved into a new studio and offices at 48 East Broadway, above the Al Buehman Photography studio. This drawing, from a promotional piece, displays the front side of the new building, adjacent to the US Post Office. When Buehman moved to a new location, his space became a carpet store, and announcers would say, "KVOA Tucson, above the Floorcraft Carpet Company on East Broadway," which helped pay the rent. When KVOA moved in 1955, its space was rented to other radio stations before being demolished to provide additional parking. (CJB.)

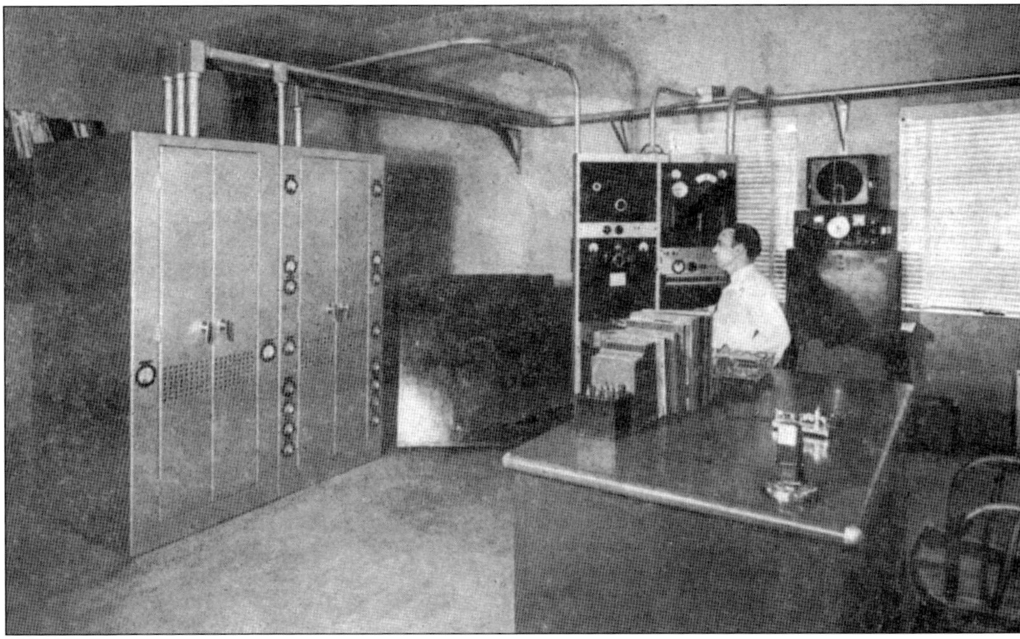

The 1940 KVOA transmitter building for 1290 AM was located on West Elm Street, where the KVOA-TV studios were built around it in 1953. The radio transmitter in the top photograph was a home-built model, which later served as the backup for a new 1-killowatt RCA transmitter that was installed next to it. In that era, FCC-licensed engineers were required to be on duty at all times when a radio station was on the air. In the photograph, engineer Ray Holsclaw keeps an eye on the equipment. The bottom photographs show the new 1940 KVOA facilities at 48 East Broadway. Announcers sat at the RCA 76B console (right) to turn on the microphone and read commercial announcements between programs, to play the records and 16-inch transcription discs that held the music for local programs, and to bring in NBC network programs. At left, an announcer can be seen through the window of the large performance studio. Groups could perform in front of the microphones and use the grand piano. It took 10 to 15 performers to do *Assignment Tucson*, a program that retold the week's news. (CJB.)

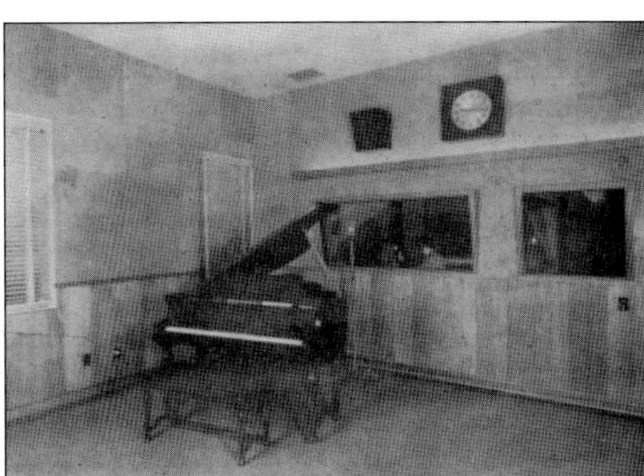

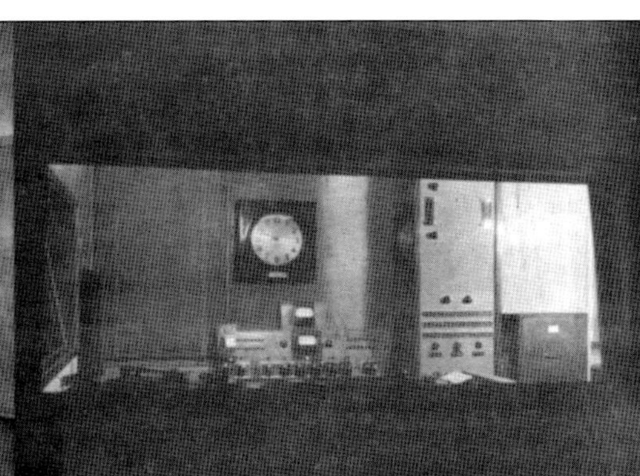

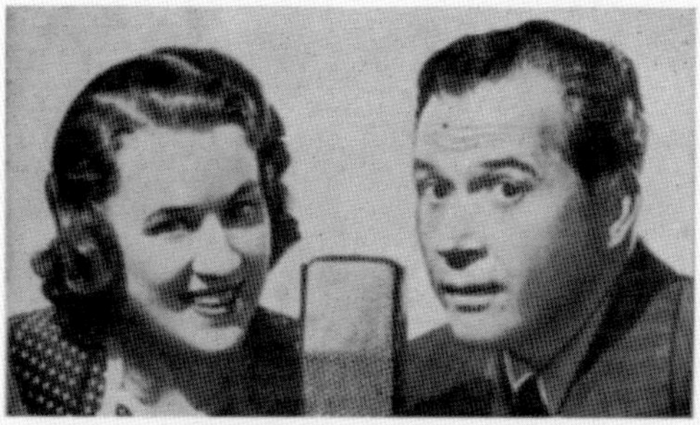

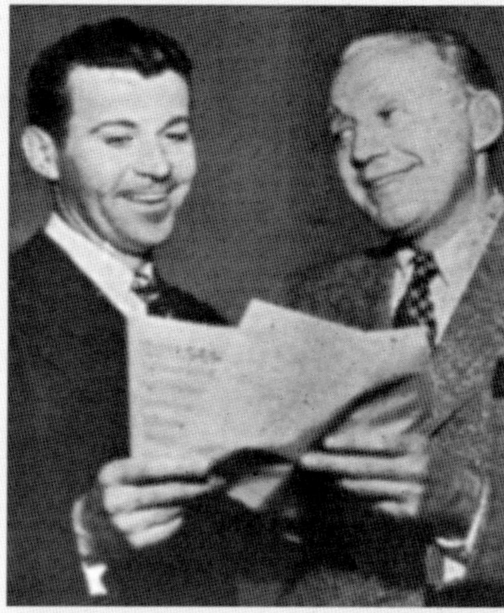

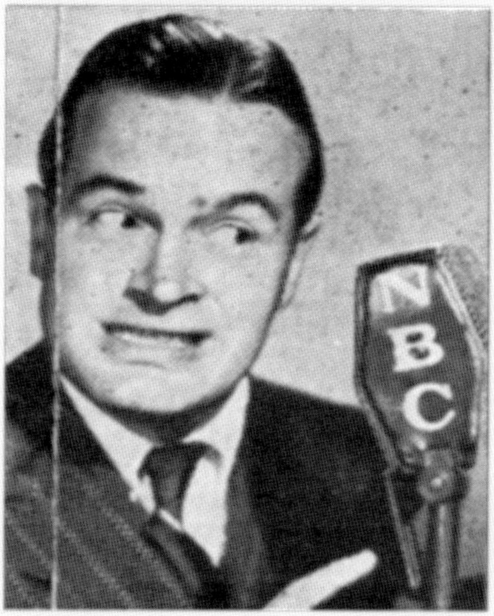

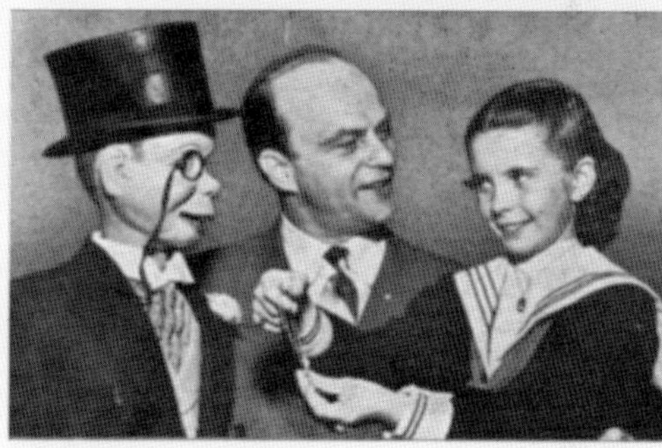

NBC had a great lineup of radio stars hosting programs that were carried on KVOA. These photographs come from a 1946 print edition of *KVOA NEWS* that was mailed to listeners. Comedian and movie star Bob Hope (lower left) told topical jokes. *Fibber McGee and Molly* (upper left) was a radio "sitcom" in which friends and neighbors came into the McGees' radio home. Listeners anticipated Fibber opening his hall closet, only to hear the sounds of everything falling out of it. The program's announcer came for a visit every week, and the sponsor's commercial would be part of the program. Jack Benny's program (upper right) anchored Sunday night radio. His stable of characters included singer Dennis Day, seen here at left with Benny. Edgar Bergen and Charlie McCarthy (lower right) proved that a ventriloquist could spark the audience's imagination. Here, they interview guest star Margaret O'Brien. (Pauline Jackson.)

TUESDAY, MAY, 1946

MORNING

4:55	News, KVOA Newsroom	
5:00	La Hora Mexicana	
6:30	La Hora Mexicana	
6:45	Mexican Serenade, Reuben's	
7:00	Yawn Patrol	
7:15	Western Roundup, Snell's Service	
7:30	News, Shamrock Dairy	
7:45	Howard Pyle, Valley Bank	ABS
8:00	Fred Waring and Pennsylvanians American Meat Institute	NBC
8:30	Words & Music, Grunewald & Adams	
8:45	David Harum, Bab-O	NBC
9:00	People's Exchange, Beaudry's	
9:15	Chuckwagon Time, Farmer's Mkt	
9:30	Lone Journey, Friskies	NBC
9:45	Maggi's Priv. Wire, Towne, Sports	NBC
10:00	News, ABS Newsroom	ABS
10:15	No Place Like Home, Reuben's	
10:30	Women In The News, Levy's	
10:45	News, Art Baker, Sierra Pine Soap	NBC
11:00	Guiding Light, Wheaties	NBC
11:15	Today's Children, Breakfast Tray	NBC
11:25	Betty Crocker, Homemaker	NBC
11:30	Women in White, Breakfast Tray	NBC
11:45	Rialto Review, Rialto Theater	

AFTERNOON

12:00	Woman of America, Dreft	NBC
12:15	Ma Perkins, Oxydol Soap	NBC
12:30	News, Steinfeld's	
12:45	Right to Happiness, Ivory Soap	NBC
1:00	Backstage Wife, Phillips	NBC
1:15	Stella Dallas, Energine	NBC
1:30	Lorenzo Jones, Bayer Aspirin	NBC
1:45	Widder Brown, Haley's M. O.	NBC
2:00	Lum 'N Abner, Alka Seltzer	ABS
2:15	Portia Faces Life, Jello Pudding	NBC
2:30	Music to Remember, Grabe's	
2:45	Meditations, Arizona Mortuary	
3:00	Ballroom Echoes, Blue Moon	
3:15	Memory Lane, Al Buehman	
3:30	Aunt Mary, Famous Flours Co.	NBC
3:45	Dr. Paul, Edward's Coffee	NBC
4:00	Heart to Heart Hour, Entrekin	ABS
4:15	News of the World	NBC
4:30	Road of Life, Duz	NBC
4:45	Joyce Jordan, MD, Dreft	NBC
5:00	Aunt Peggy	
5:15	Hawaiian Melodies, Wallis Cleaners	
5:30	News, Porter's	
5:45	The Town Crier	
5:55	Latest Sports Summary	

EVENING

6:00	Amos 'N Andy Rinso	NBC
6:30	Fibber McGee & Molly Johnson Wax	NBC
7:00	Bob Hope Pepsodent	NBC
7:30	Red Skelton Raleigh Cigarettes	NBC
8:00	Supper Club, Chesterfield	NBC
8:15	News, Tucson Laundry	
8:30	Frolics of 1946 Philip Morris	NBC
9:00	The Shadow Baum & Adamson	
9:30	Date With Judy Tums	NBC
10:00	Music Pro and Con Symposium in Jazz	
10:30	News, Richfield Reporter	NBC
10:45	Fun Time, La Jolla	
11:00	Music By Request Tucson's Musical Favorites	
11:55	News, KVOA Newsroom	
12:00	Buenos Noches, Amigos de Tucson, de KVOA	

This is the KVOA Tuesday radio program schedule in April 1946, taken from *KVOA NEWS*, just as Arizona was to observe daylight savings time. Primetime Tuesday evening programs from NBC featured some of the biggest comedy stars of the years before and after World War II. The 15-minute newscasts originated at KTAR in Phoenix, but note the 4:15 p.m. NBC News, with Morgan Beatty, which was the predecessor for the network's television *Nightly News*. Program sponsors included some that are still in business, but the cigarette companies are long gone from the airwaves. NBC's 15-minute afternoon programs were radio soap operas, each one telling a continuing story and usually sponsored by a soap manufacturer. KVOA mornings started at 5:00 with *La Hora Mexicana* and host Don Jacinto Orozco. *Arizona Highlights* with Howard Pyle, broadcast from Phoenix, covered statewide news and interviews. Pyle left broadcasting and was elected governor of Arizona in 1950. *Chuckwagon Time* was a Western music program hosted by rotund announcer Jean Lely, who dressed in Western apparel for personal appearances at the Saturday-night KVOA Barn Dance. From 2:30 p.m. to 3:30 p.m., the 15-minute music programs were each hosted by a different announcer and had a separate local sponsor. The *Town Crier* was a local interview program hosted by Ben Slack. The day concluded with *Music By Request*, which played popular records from the big band era by artists such as Glenn Miller and Tommy Dorsey, and songs from Frank Sinatra and Bing Crosby. (Pauline Jackson.)

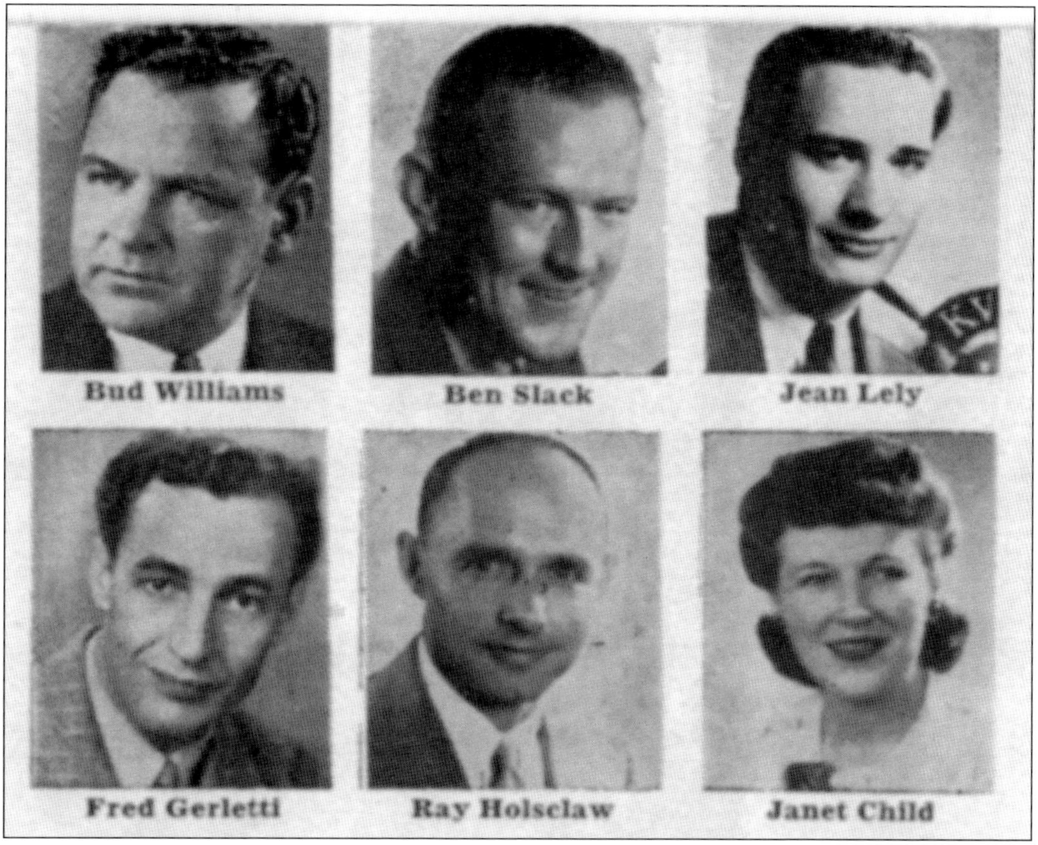

Some of the KVOA radio executives and 22-person staff are seen here in the 1946 *KVOA NEWS*. R.B. "Bud" Williams was the general manager who led the station through the postwar era and eventually into television. Ben Slack was the program director, in charge of the on-air staff hiring and utilization. KVOA alumnus Larry Schnebly remembers his first day at KVOA, in 1953, when Slack pointed to a plaque above the clock that read, "If there is any doubt about what to DO, what to SAY or what to PLAY, DON'T DO IT . . . DON'T SAY IT . . . DON'T PLAY IT." (Pauline Jackson.)

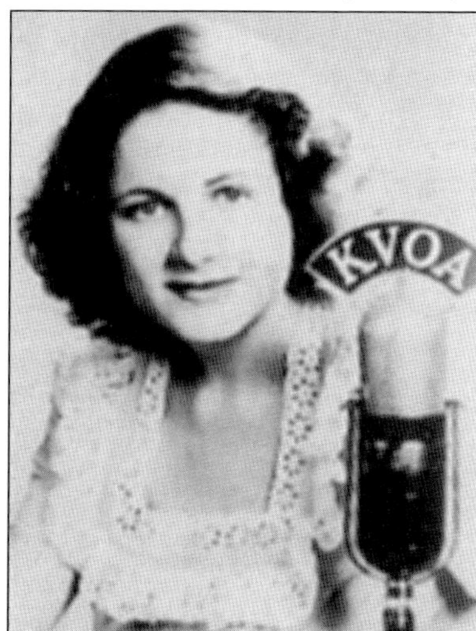

Priscilla Sanders, an announcer at KVOA starting in the 1940s, was one of the first females to hold that job. She worked afternoons doing local announcements between soap operas and writing scripts for her programs, as most announcers did. This photograph is from a "personal postcard" from 1948 that was saved by her daughter Deborah Manspeaker Weiss. Sanders's family heritage went back to the pioneer days; her great-uncle was Monte Mansfeld, Tucson's Ford dealer. She was married to Irving Manspeaker, whom she met at KVOA. Their descendants have remained as Tucsonans for generations. (Rob Weiss, OP Coins.)

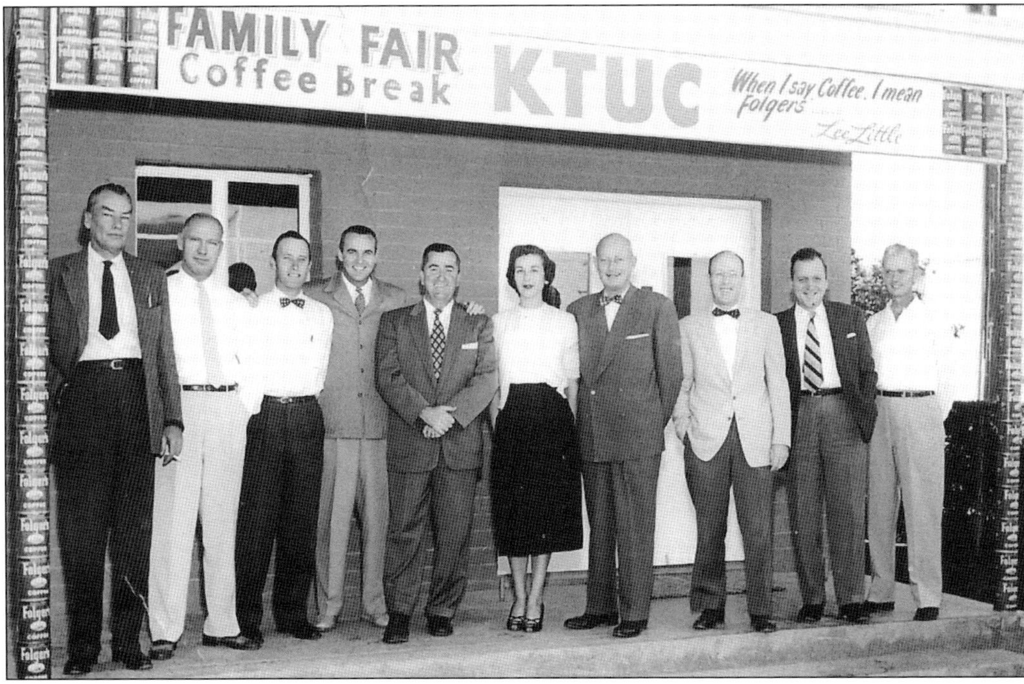

The KTUC call letters were assigned to 1400 on the dial in 1940. The station moved from the KGAR location to a new building at the corner of Broadway and Tyndall Avenue in 1942. After the war, Monte Mansfeld Motors built its dealership across the street. Around 1950, the group shown above gathered on the front porch to kick off a promotion for sponsor Folgers Coffee. Members of the early KTUC staff are pictured, including newscaster Fred Finney (far left), sales manager and morning show cohost George Bradley (second from left), and Hilde Clark (sixth from left), who did "women's" programs. She is standing next to Lee Little (seventh from left), the station owner, manager, and host of his own popular morning program. Pete Tufts (second from right) was a station announcer and the program director. The others are presumed to be sponsor representatives. Through its affiliation with CBS, the station carried network programs, including Sunday concerts by the New York Philharmonic and, daily, three hours of 15-minute CBS soap operas, including *The Guiding Light*, which began on radio and moved into a long run on CBS television. CBS newscasts were done by Edward R. Murrow's team, which built the great CBS News department. In 1938, the team instituted the *CBS World News Roundup*, featuring live reports from correspondents around the world. The roundup was still being heard daily 75 years later. (Above, MaryLee Carter; below, CJB.)

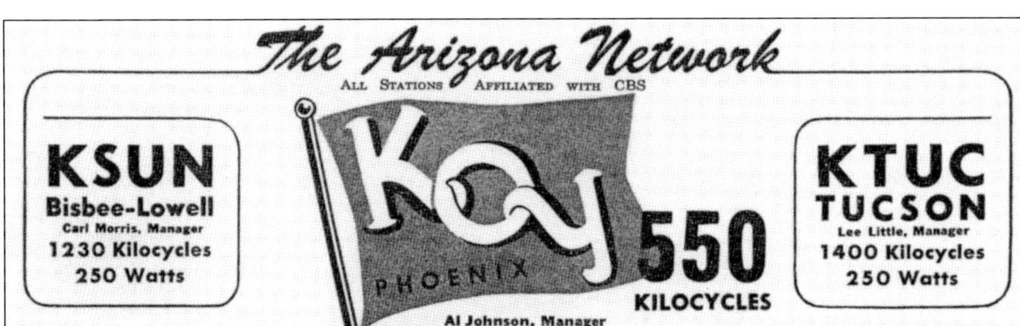

George Bradley (standing) and John "Square Deal" McNeil teamed up to host Tucson's popular morning program on KTUC (and briefly on KTAN) from the late 1940s to the 1960s. Bradley was also the station sales manager, having gone into radio as an alternative to climbing telephone company poles. McNeil was a car salesman at Central Auto Sales on South Stone Avenue when it became the sponsor of a 15-minute program, which grew into a two-hour show. Listeners enjoyed the banter between the partners, as Bradley was a mischievous jokester and McNeil knew how to react. (MaryLee Carter.)

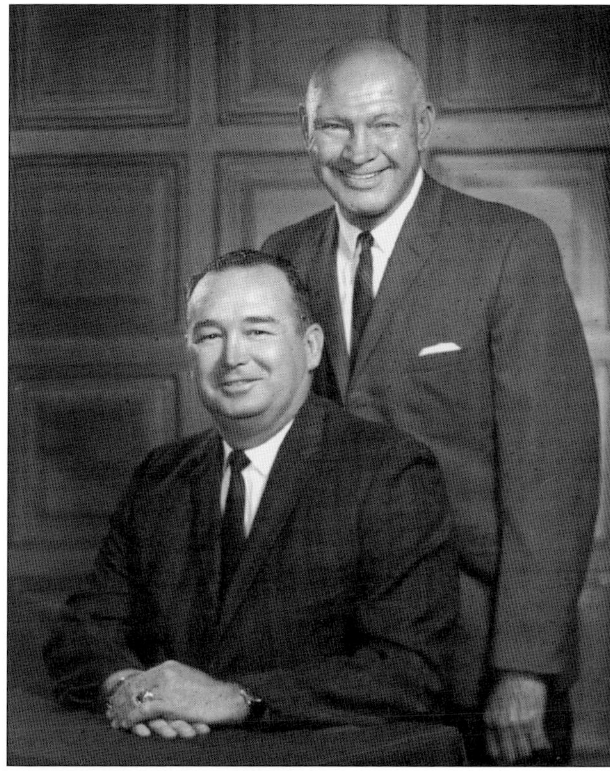

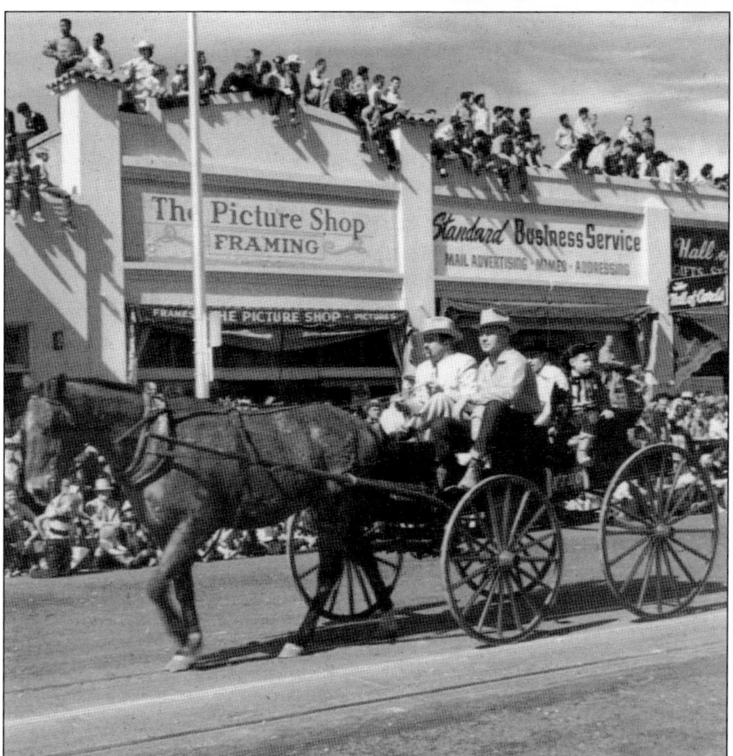

George Bradley (right) and "Square Deal" McNeil rode in KTUC's buggy in the Tucson Rodeo Parade. Bradley's family remembers the benefit of having their dad selling advertising. His daughter, MaryLee Carter, says, "At Christmas we always were treated to gifts from his advertisers, who definitely wanted to keep him on their good side so that he would mention them on the morning show. . . . One Christmas we were inundated with candy, so much so that we kids grew tired of it and cheered when Dad told us that the five-pound box of candy he left in his car had been stolen!" (MaryLee Carter.)

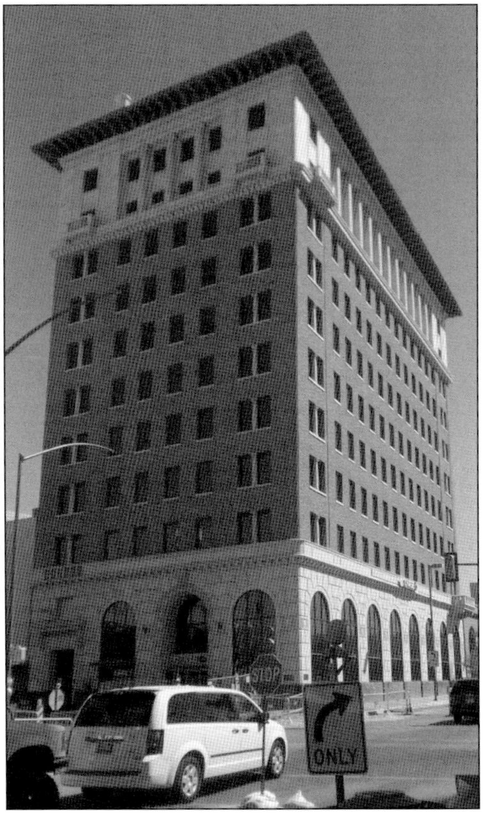

Standing at the prestigious corner of Congress Street and Stone Avenue, the Valley National Bank has been one of Tucson's tallest buildings since the 1930s. KVOA had its rooftop transmitter here before it moved to Elm Street. The bank sponsored radio programs, including a "man on the street" feature, wherein announcers interviewed people passing by the front of the building, called *Hank and Frank at the Valley Bank*. Frank was Frank Barreca, a KVOA radio announcer who went into television and was with KUAT-TV for many years before going into the travel business. Hank was Hank Hubbard (below), who was one of the most popular broadcasters in Tucson for decades. He was an announcer for KVOA radio before moving to its television station in 1953. He did many radio shows, including *Assignment Tucson*, but developed into a television personality as a commercial announcer and weatherman. His career skyrocketed once he became a news anchorman on channels 4 and 13. (Left, RJ; below, Pauline Jackson.)

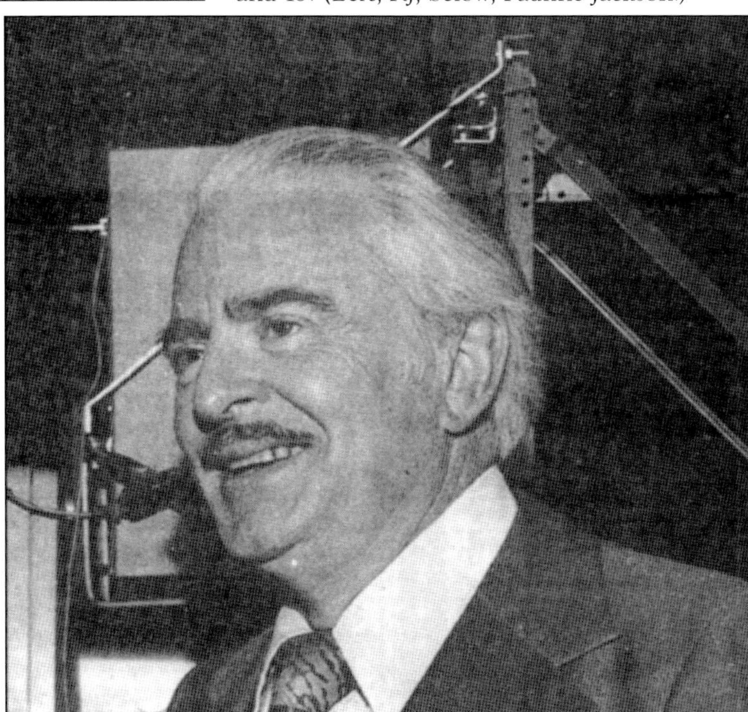

Two

After World War II

New Stations

While KTUC and KVOA were serving the community well, there was room for more stations to compete for the ears of the ever-expanding Tucson population. KOPO, at 1450 on the dial, became the first new station, signing on at 6:00 a.m. on February 27, 1947. That same year, KCNA joined the airwaves at 1340 on April 10.

These four stations had the Tucson airwaves to themselves until December 13, 1949, when KTKT signed on at 1490. KCNA made the move in November 1951 to 580, the opposite end of the dial, making it difficult for listeners to crank their dials to hear it. KEVT, an all-Spanish station, signed on in September 1953 at 690. FM radio was in its infancy nationwide, but in March 1954, KTKT-FM took to the air with its classical music format at 99.5 on the otherwise empty FM dial. KTKT-AM wanted the wider coverage that a move to 990 would provide with a big 10-kilowatt signal, but reverted to operating in the daytime only. On the day KTKT left 1490 in January 1957, KAIR immediately came on the air. With KMOP taking 1330 in 1957, and KCEE on 790 in 1958, the AM band was quite competitive for Tucson broadcasters into the late 1950s.

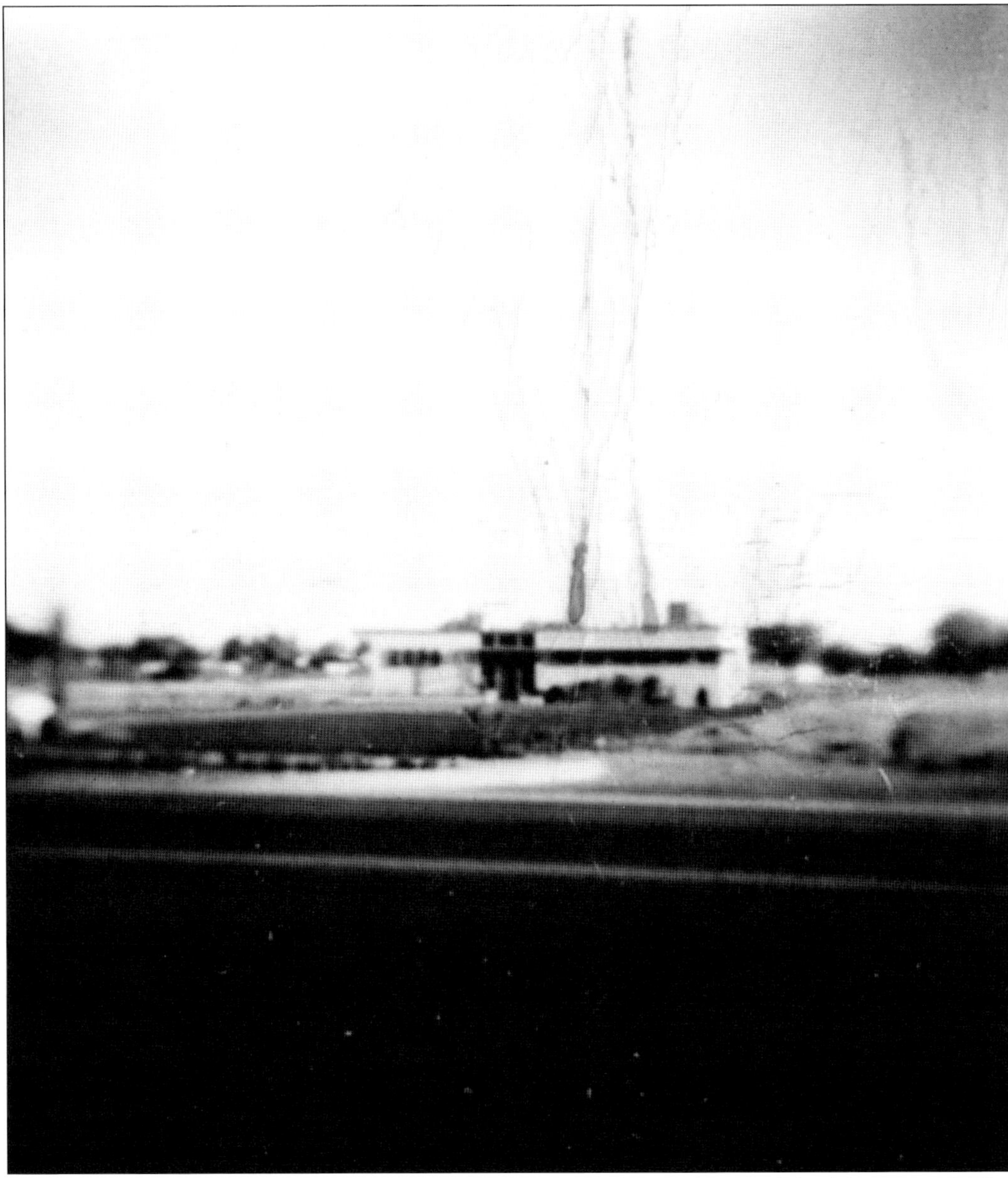

When KOPO came on the air in 1947 at 1450, it operated from this small building at 115 West Drachman Street. The transmitter tower can be seen behind the building. In 1952, that long lawn was dug up, and a large building was erected to house the new KOPO-TV, channel 13. KOPO radio, the first Tucson affiliate of the ABC radio network, carried programs like Don McNeill's *Breakfast Club*, a variety hour heard every morning at 8:00 live from Chicago. Local programs on KOPO radio included a midday country-western music show hosted by Buster "Bus" Williams, with engineer Vern "Saddlesore" Cimmiyotti. After 10:00 p.m., announcer Matt Lemen hosted *Tempo Rendezvous*, playing popular songs of the day and talking about the great artists he was playing. Lemen also hosted special programs, like a 1949 half-hour live show with the Sons of the Pioneers before the group appeared in concert later that night. Lemen left broadcasting and worked as an employment interviewer with the State of Arizona, but retained his interest in music and radio. (RJ.)

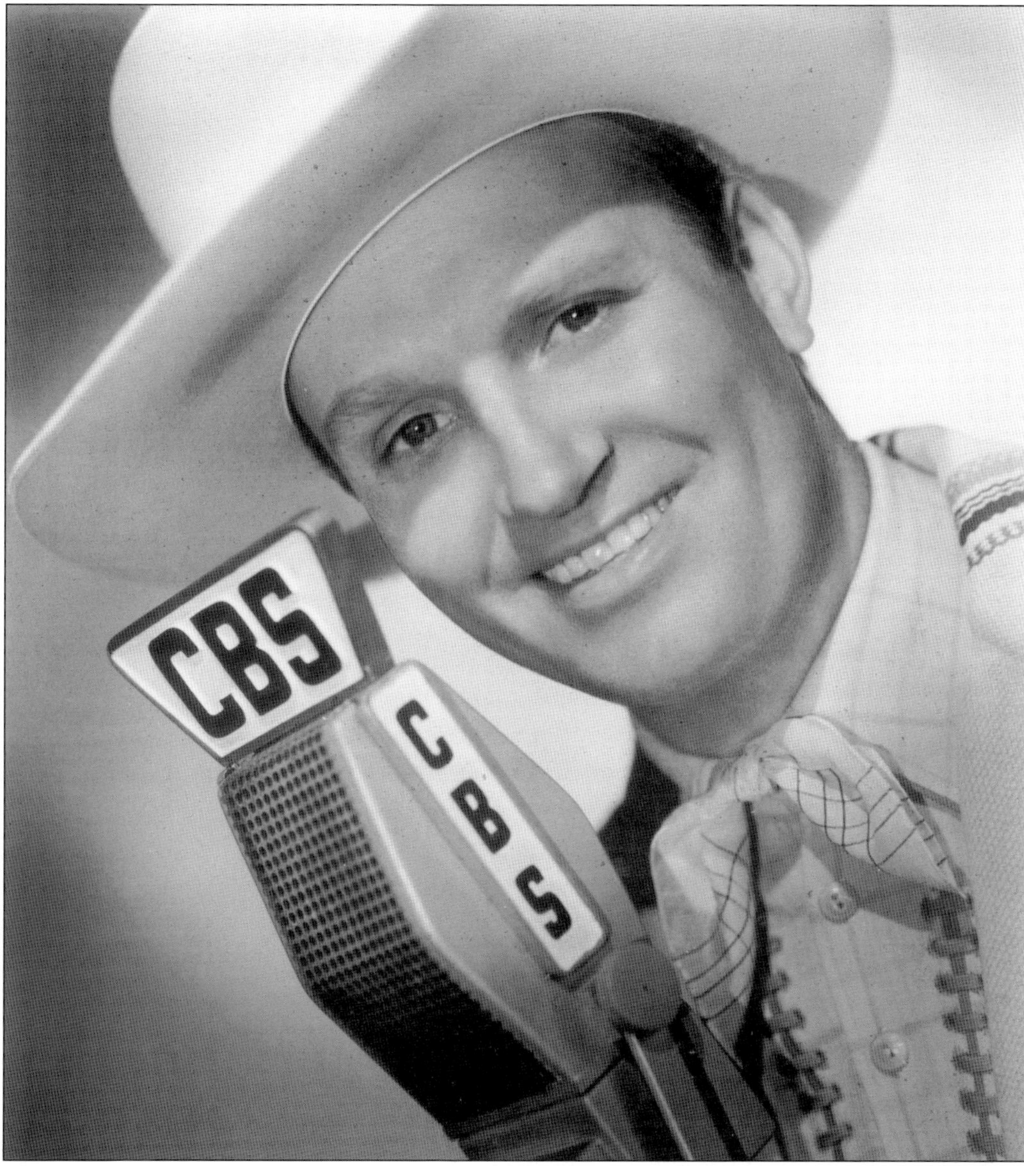

Gene Autry was a major Western movie star. On his popular CBS radio program, he starred in stories of law and order in the Old West and sang his songs. Those songs were also on records, and his two biggest hits were his theme song, "Back in the Saddle Again," and the all-time seasonal favorite, "Rudolph the Red-Nosed Reindeer." Autry's personal appearances with his horse, Champion, drew huge crowds. He was also a very successful businessman. He was the original owner of the California Angels baseball team, and he invested in radio and television stations, including Tucson's KOPO, with his partner, Phoenix broadcaster Tom Chauncey. The two acquired the CBS radio network affiliation for their stations, ensuring that Autry's program would be heard on the stations he owned! That switch took place on January 1, 1950. In 1953, Autry's television show was seen on Tucson's first television station, which Autry and Chauncey owned, KOPO-TV. The station has remained a CBS affiliate for decades. Radio personality Chris Borden remembers working for Gene Autry at KSFO in San Francisco, "He was a wonderful guy and treated all of his employees like old friends." (Autry Trust.)

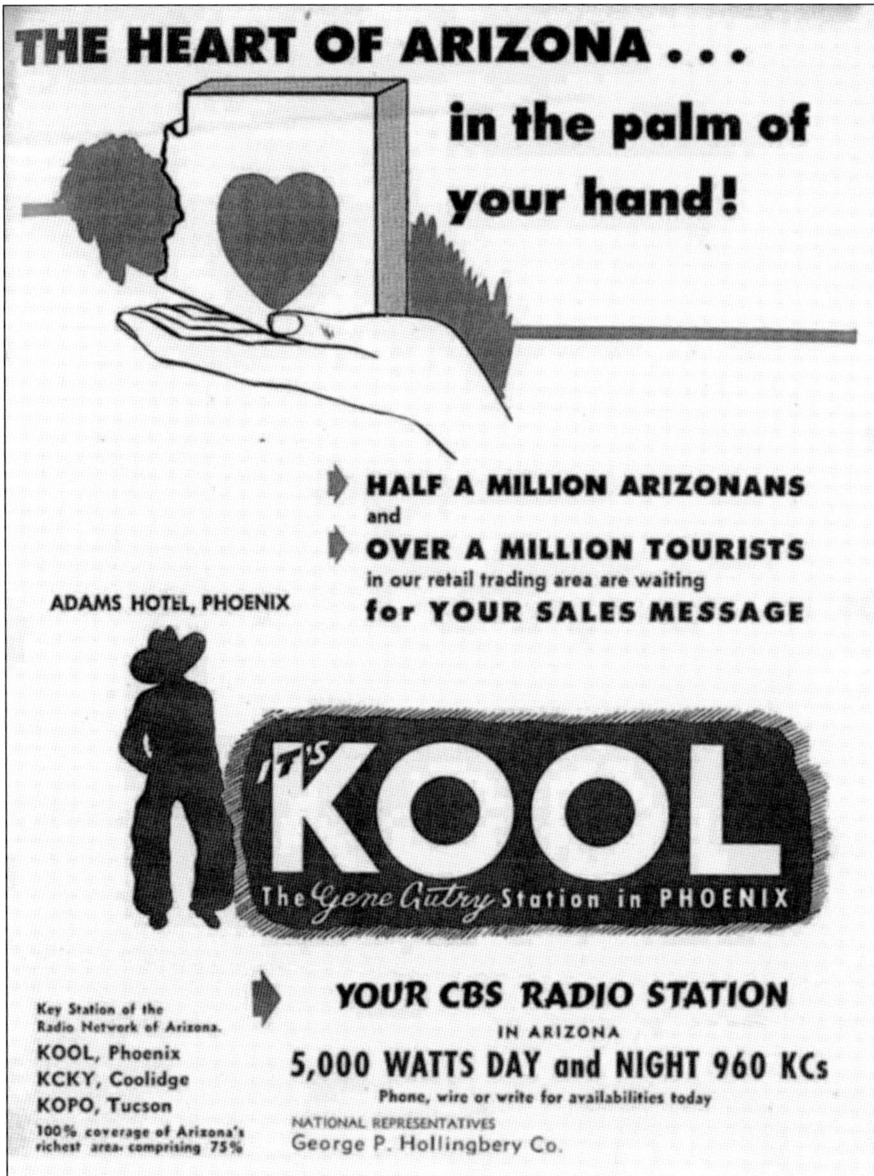

Gene Autry and Tom Chauncey's company owned both KOPO in Tucson and KOOL 960 in Phoenix. The call letters of KOPO were changed to KOLD in the 1950s. When KOPO became a CBS affiliate, it acquired the network's dramas, Westerns, comedies, and soap operas at a time when network radio was near its popular peak. New radio stars, like Arthur Godfrey and Art Linkletter (*House Party*), were on CBS radio daytime, and the evening programs included *My Favorite Husband* with Lucille Ball, *Suspense*, *Gunsmoke*, and Stan Freberg's show. Network radio programs ended by 1960, but news and sports continued. Locally, KOPO had a staff that remained with the station into the television era. E.S. Mittendorf was the general manager, and his wife, Virginia, worked in the office. She became a local star with her live afternoon interview hour with sidekick Gerry O'Brien and organist Dude Vance. Sales manager Paul Plunkett was the television station's weatherman. Salesman and sportscaster Bernie Perlin covered live high school and University of Arizona (UA) sports events as well as the Tucson Rodeo. (CJB.)

KCNA signed on at 1340 in 1947 as an independent station with no network affiliation. The studios, built at Sixteenth Street and South Cherry Avenue, used the self-supporting tower at right. The morning program was called *The 3 T's* (time, tunes, and temperature). The Catalina Broadcasting Co. ownership included the Chambers family from the *Tucson Citizen* and the author Erskine Caldwell. Many of the original staff were hired away from KTUC, including Wayne Sanders, who became station manager, Ray Helgeson as program director, and announcers Jim Officer and Al Morris. In July 1948, management cited demand from advertising clients to have network programs to hang their commercials around. The station became Tucson's first affiliate of the Mutual Broadcasting System (MBS), which cut the amount of local programming considerably. Added were MBS network newscaster-commentators Frank Hemmingway, Fulton Lewis Jr., and Gabriel Heatter; mysteries *The Shadow* and *Quiet Please*; and game shows *Queen for a Day* and *20 Questions*. Morris returned in 1949 to start a new KCNA program, the *Arizona Hayride*, featuring country-western music. The popular music programs were *1340 Club* and *Best by Request*, with host Jim Officer. The arrival of KCNA was announced in the advertisement below. (CJB.)

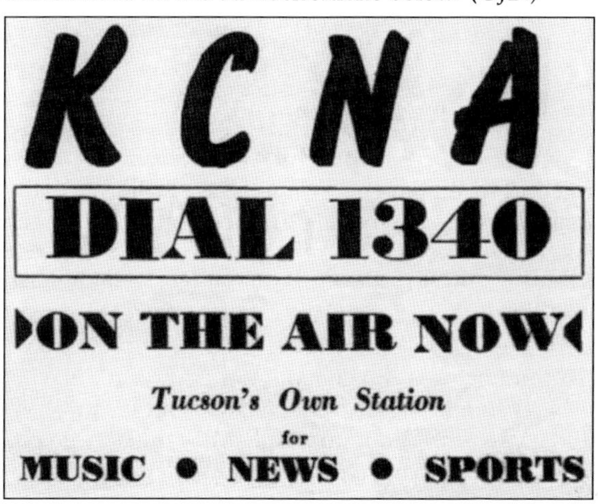

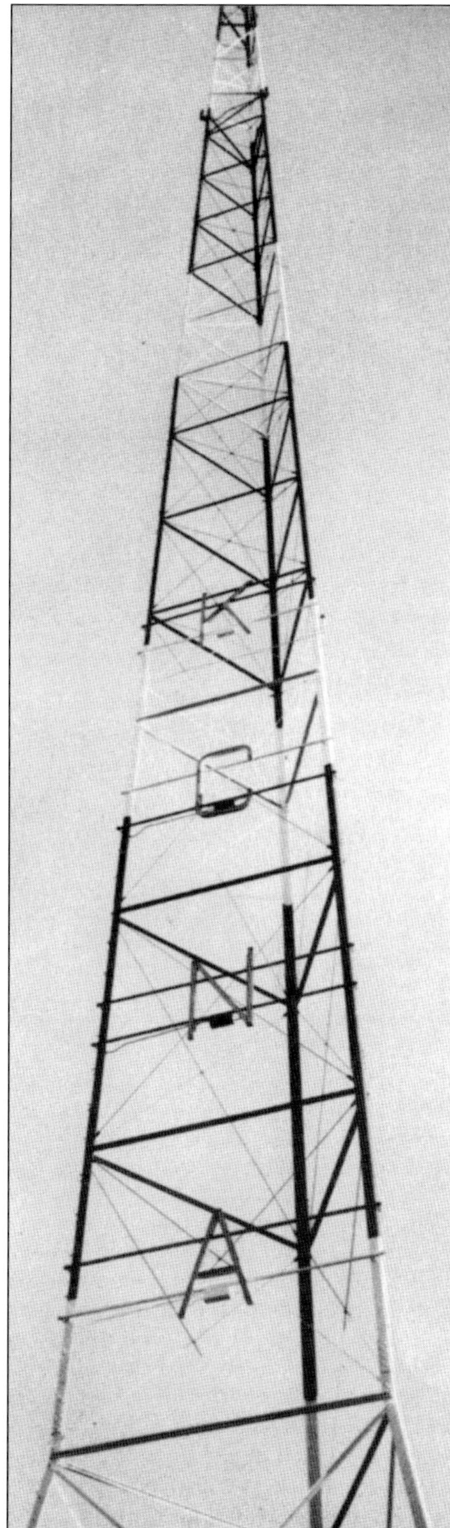

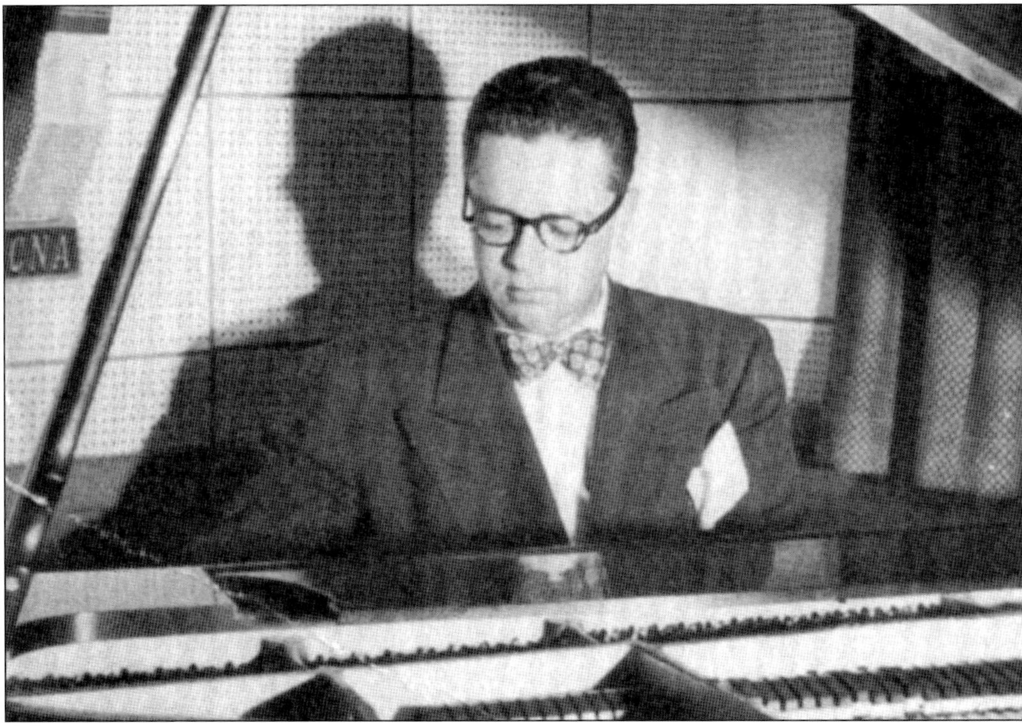

Wayne Sanders, KCNA's general manager, was a very talented musician. His career included playing piano in dance bands, and he started his own radio program when he was KTUC program director. That program came with him to KCNA, where he made sure the studio had the grand piano shown here. The station had a variety of musical programs as well as community-service group interviews and sportscasts. Sanders continued to play piano around Tucson. This and the other KCNA-related photographs are from a 1949 promotional brochure. (CJB.)

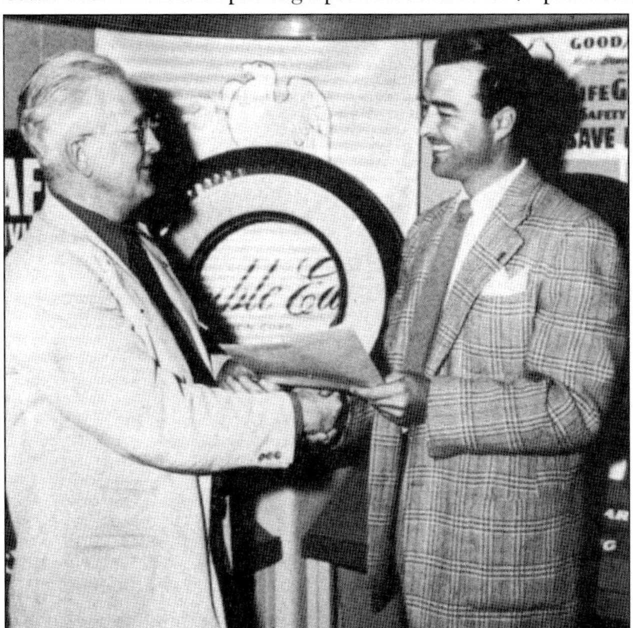

Harry Chambers (right) returned from World War II service and joined the sales staff of KCNA, which his family partly owned. Nicknamed "Handsome Harry," he had taken a screen test in Hollywood, but he preferred to be home in Tucson. Here, he shakes hands with Tom Talmadge of Apache Tire after the company signed a contract to sponsor the 7:30 a.m. newscast, Monday through Saturday. Chambers became sales manager, then general manager, and was the station's primary owner until it was sold in 1967. Engineer Gene Stough joined the staff in 1950 and stayed with Chambers because of the personal relationships he developed. (CJB.)

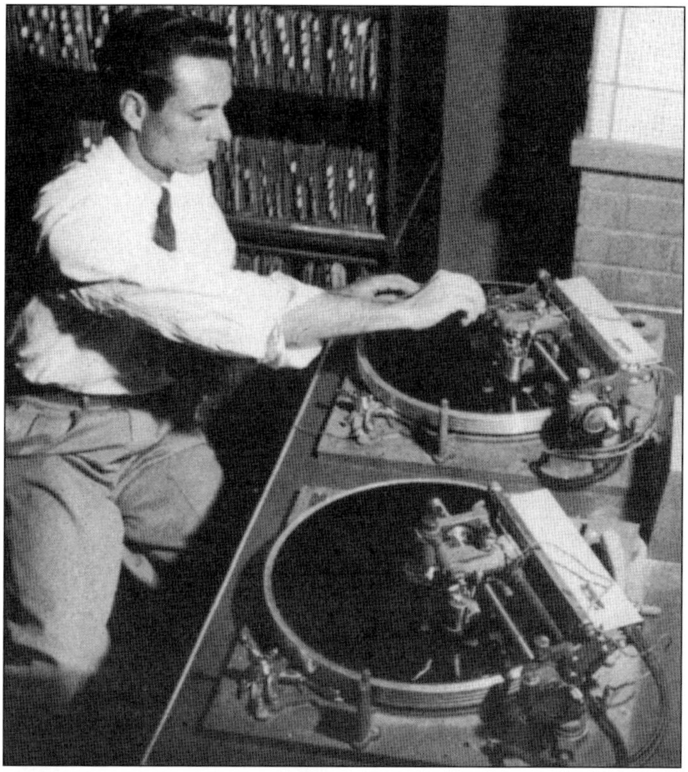

KCNA engineer Tom Epperson tends to the 16-inch disc cutters that the station used to record programs and announcements for later broadcast. Tape recordings would not become available until the late 1940s and were not easily used to record short announcements, so stations continued to "cut" commercials and other short items until self-cueing tape cartridges were introduced in the late 1950s. A 16-inch disc could hold only 15 minutes of material, and no editing was possible. If a mistake was made, a new disc was necessary. (CJB.)

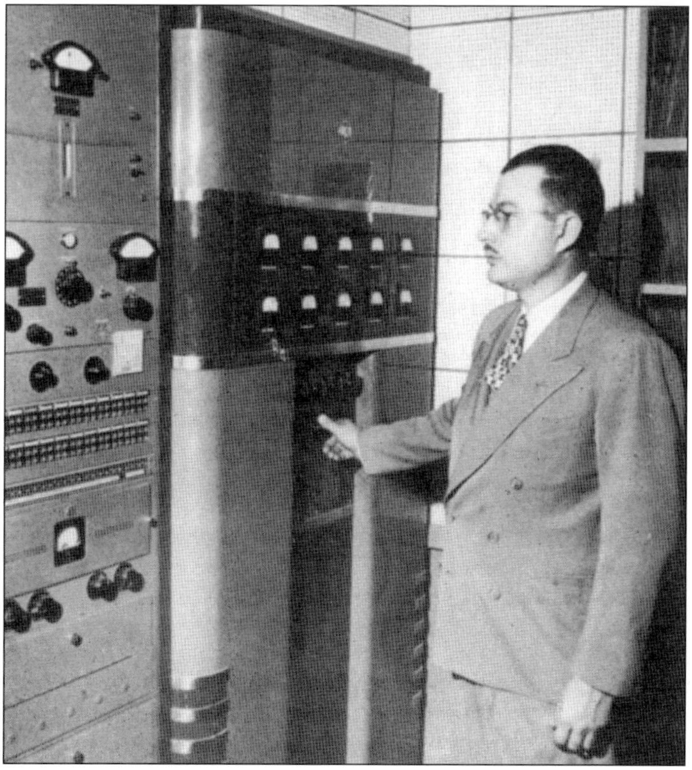

KCNA chief engineer Bill Karpisek built the physical facility using RCA equipment. Here, he tunes the RCA 250-watt transmitter for 1340 AM. That transmitter was the size of a modern-day refrigerator. In the 1940s and 1950s, an engineer with a first-class license was required to be on duty wherever a broadcast transmitter was located and whenever it was on the air, in order to control its operation. Karpisek's staff included Kenny Blue and Ken Ferguson, in addition to Tom Epperson. Engineers also operated the control board, which freed the announcers to announce. Bill Karpisek left Tucson for Sacramento, where he was chief engineer for KCRA-TV. (CJB.)

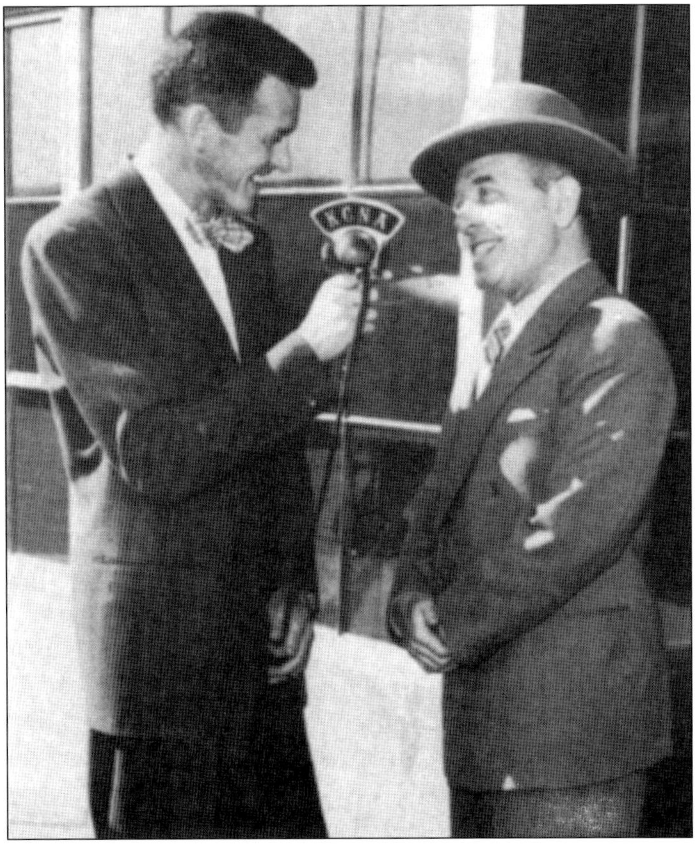

Announcer Bob Keyworth (left) took his microphone to the Tucson Southern Pacific train depot for this interview with Eugene Ormandy, the conductor of the Philadelphia Orchestra. There were classical music programs on KCNA, and the station carried the *Sunday Evening Forum*. Keyworth did several programs on KCNA, including *Man About Town* and the audience-participation show *Luncheon in Tucson*. On Saturday mornings, he did *Flight Talk*, about the rapidly growing aviation industry. Each week, a caller could win a free flight over Tucson in a small airplane. (CJB.)

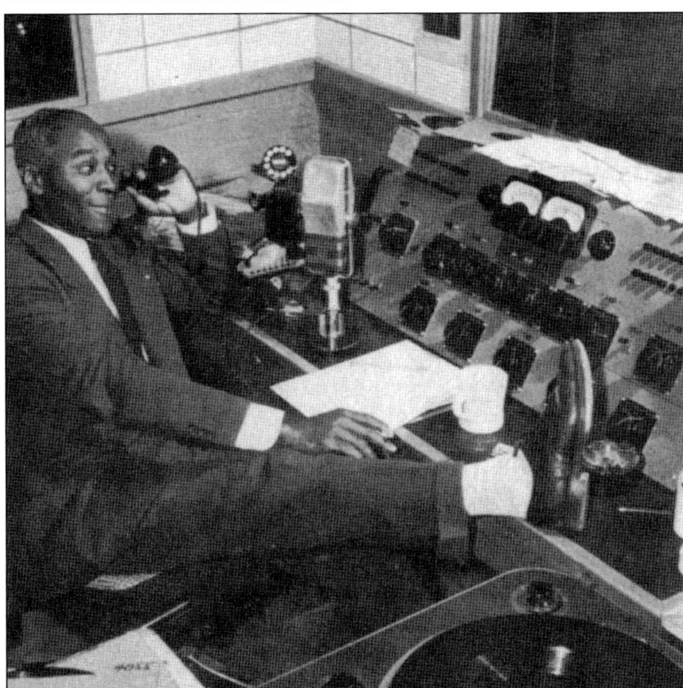

The original schedule of KCNA programs had the station on the air from 7:00 a.m. to 11:00 p.m. Monday through Saturday and 8:00 a.m. to 10:00 p.m. on Sundays. The first expansion of that schedule was for the *Harlem Jamboree* hour, which aired nightly at 11:00. The announcer/record spinner was Jim Brownlow, the first African American to be heard on Tucson radio. Brownlow, formerly a trombonist with the great singer Bessie Smith, was the station's contract janitor. When manager Sanders learned about his musical background, Brownlow was quickly invited to go on the air. (CJB.)

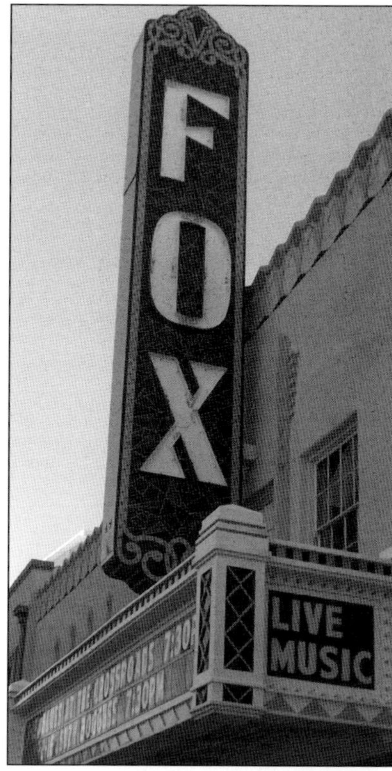

Radio programs for kids included not only network offerings like *Superman*, *Terry and the Pirates*, *No School Today*, and *Let's Pretend*; in Tucson on every Saturday morning in the early 1950s, KCNA broadcast a live program from the stage of the Fox Theater downtown. Bob McKeehan gave contest prizes to kids attending the Mickey Mouse Club movies. KCNA also broadcast a live talent show, *Talent Quest*, with host Wayne Sanders from that stage. (RJ.)

Students from Catalina Jr. High (now B.C. Doolen Middle School) participate in a Saturday morning show on KCNA. The *Peter Grose Show*, featuring fun and music, was on from 1949 to 1950. Peter (center) is joined by his staff of "young irrepressibles." They are, from left to right, Larry Becker, Larry Leahy, Jack Rosensweig, and Russ Jackson. Becker became a minister, Leahy joined the Army, Rosensweig stayed with the station for years before becoming an accountant in California, and Jackson is the coauthor of this book, following a career in radio and teaching. Peter Grose graduated from Yale and Oxford, worked as a journalist for the Associated Press and the *New York Times*, and wrote several books. (CJB.)

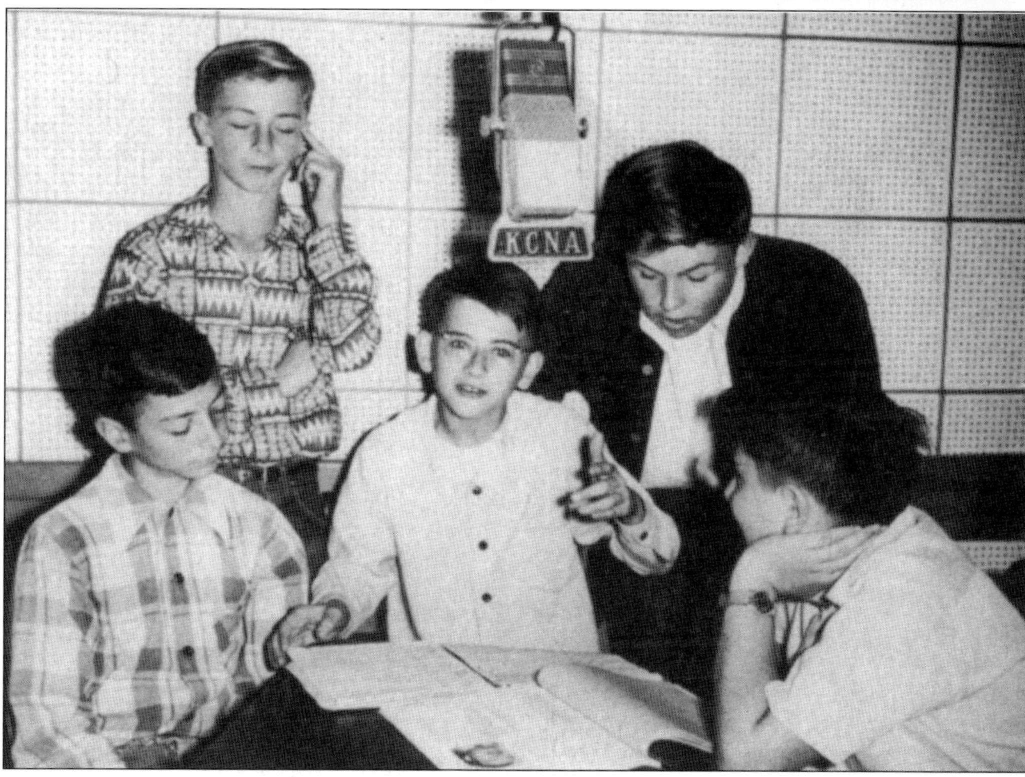

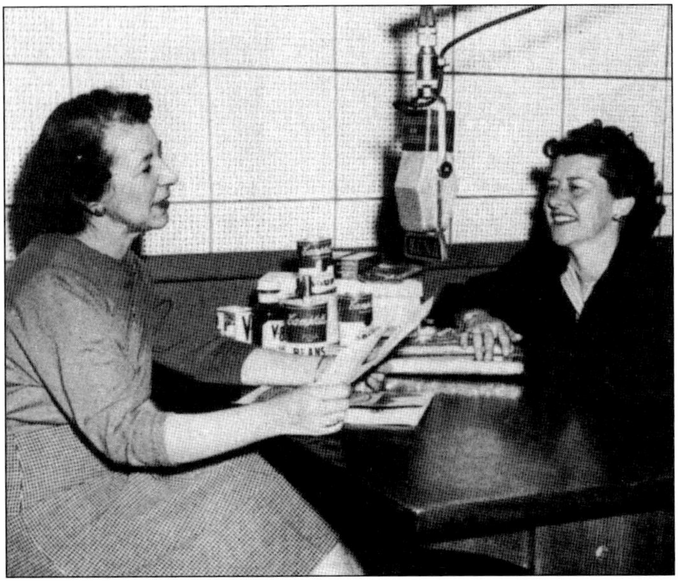

Programs for women were scheduled on local stations in the daytime, as most women were housewives, busy fulfilling domestic duties and familial obligations. They became the target market for sponsors. These radio programs addressed topics like fashion and food. Here, KCNA's Freda Botkin (left) interviews Ruth Metz, Tucson's Welcome Wagon director, about gift-giving etiquette. Botkin was always looking for bargains, interesting people to interview, and special events she could share with listeners. (CJB.)

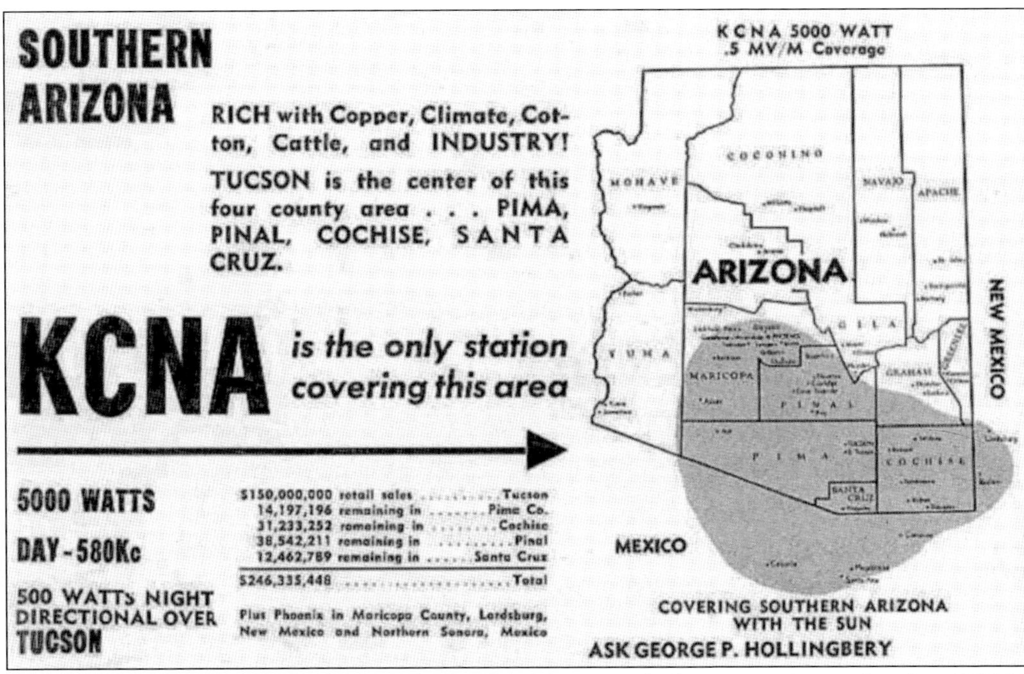

The years 1950 and 1951 saw changes for KCNA. First was the big network switch, when KCNA became an affiliate of ABC, KTUC went to MBS, and KOPO moved to CBS. On November 8, 1951, KCNA moved its dial position to 580. Its 5,000 watts covered all of southern Arizona, as seen in this coverage map and station advertisement. KCNA's programs were now available to listeners who could not have picked up the station's 250-watt signal when it was at 1340. The new signal brought programming to the then Papago Indian Reservation west of Tucson, where its residents loved the *Arizona Hayride* program of country-western music. (CJB.)

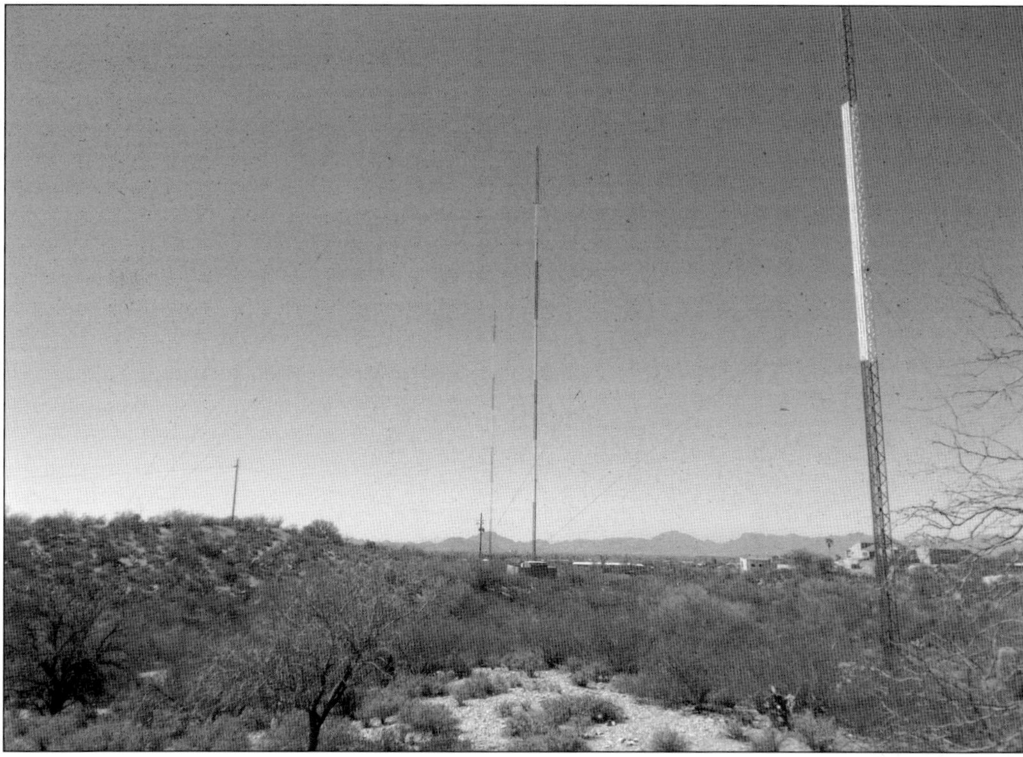

The new KCNA transmitter for 580 was located on property owned by Erskine Caldwell, north of River Road at Swan Road. These three towers were constructed there, over the objection of a photographer whose home was nearby. He thought the steel towers would obstruct his view of the Catalina Mountains, and he came out with his shotgun to try to stop the construction. A sheriff's deputy was summoned, and he told the photographer that the sheriff would see him downtown. The deputy then told the crew that the road had better be done in two hours. The towers, still standing, are today owned by the city. (RJ.)

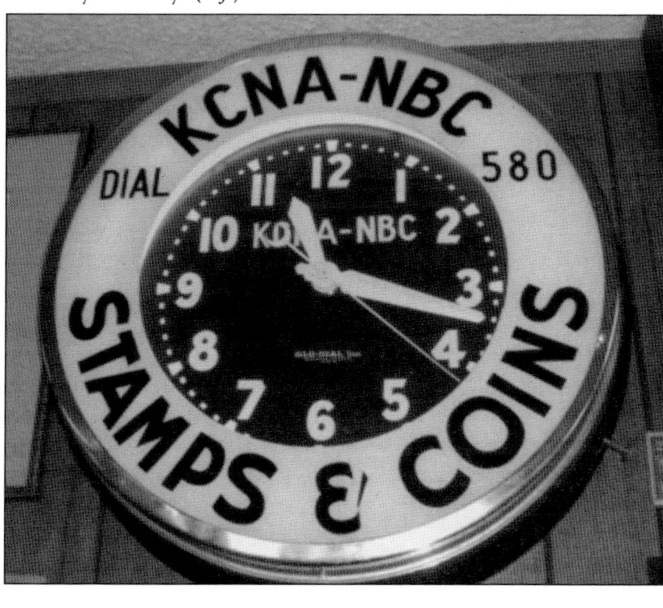

In 1956, ownership of KCNA passed to Harry Chambers, and he moved the studios to a new building at the Sands Hotel, at Congress Street and the freeway. In 1957, Chambers dropped ABC and went to a format of pop music, ending the *Arizona Hayride*. In 1958, KCNA became an affiliate of NBC after KCUB's new owners dropped it. This clock is located today at OP Coins on East Speedway Boulevard. In 1959, KCNA changed its call letters to KTAN. The station was sold in 1967, and Chambers went into banking. (Rob Weiss.)

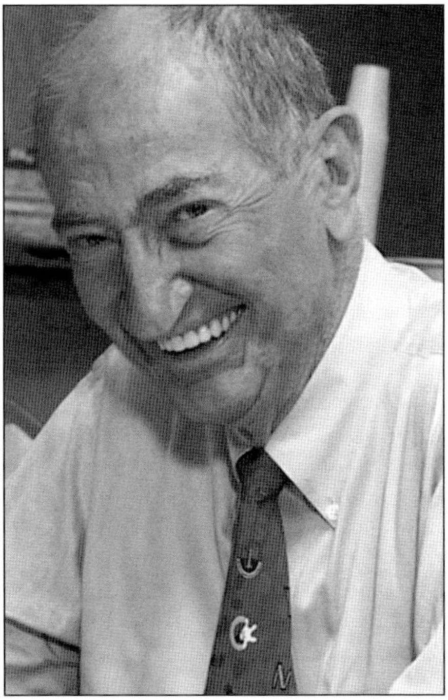

The friendly voice of Larry Schnebly (left) is recognizable to Tucson audiences from his time on KGUN-TV and his frequent hosting gigs around town. Schnebly came to Tucson in September 1953 as an announcer for KVOA radio, and he and his wife Lee have lived in Tucson, in the same house, ever since. The Schnebly connection to Arizona goes back to the late 1800s, when the town of Sedona was named after Larry's grandmother, Sedona Schnebly. The Schnebly family has been called Tucson's first family of the media. Lee, a teacher and author, hosted a Saturday-night record program on KCNA, playing the studio's grand piano and talking about the songs. Daughter Laurie works in advertising in Phoenix and has written several books. Daughter Lisa Schnebly Heidinger is also an accomplished author. Lindsay worked at Disneyland as head of the publicity department and owns a production business. Lyle, the youngest Schnebly, works in the voiceover business in California. As KVOA radio's program director, Schnebly helped construct the new radio facility on the television site (below) in 1955. (Left, Larry Schnebly; below, RJ.)

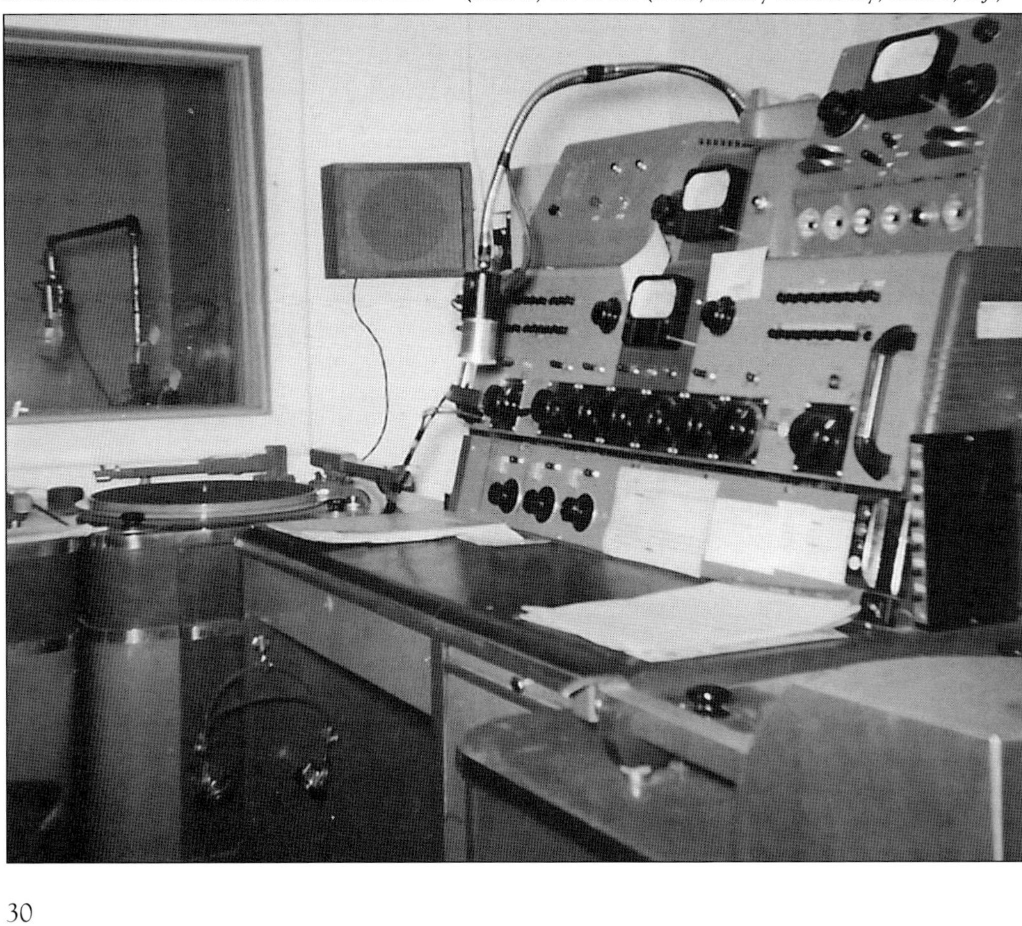

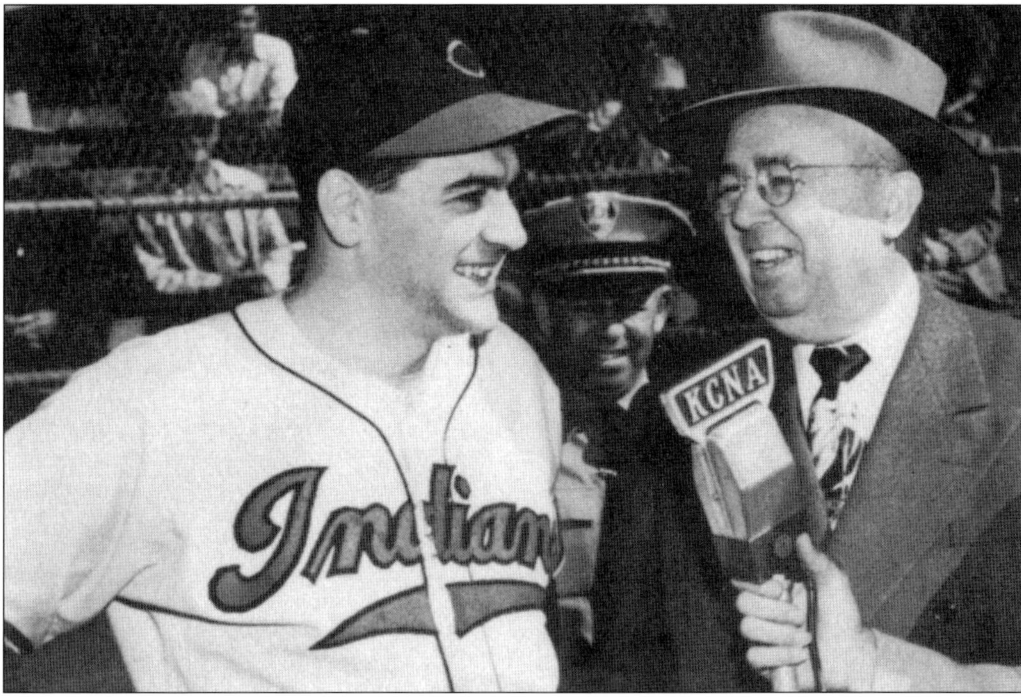

Sports on radio dates to nearly the beginning of the medium, with coverage of football, baseball, and boxing. Here, Arizona governor Dan Garvey welcomes Cleveland Indians manager-shortstop Lou Boudreau in 1949. Stan Norman, Jim Officer, Al Lurie, Pete Tufts, Bill Rider, Ray McNally, Ralph Anderson, and Bernie Perlin called live games. Chris Borden remembers Tucson Cowboys baseball road game "recreations" when he was an engineer for KTUC. "The AP wire score feeds sometimes got cut off, so Al [Lurie] used to ad lib all sorts of excuses to drag out the game, i.e. a 'player injury,' 'the lights went out,' 'a stray dog on the field,' etc., and he used wooden sticks to make the sound of the ball being hit while I had big transcriptions playing crowd noises behind his play-by-play." (CJB.)

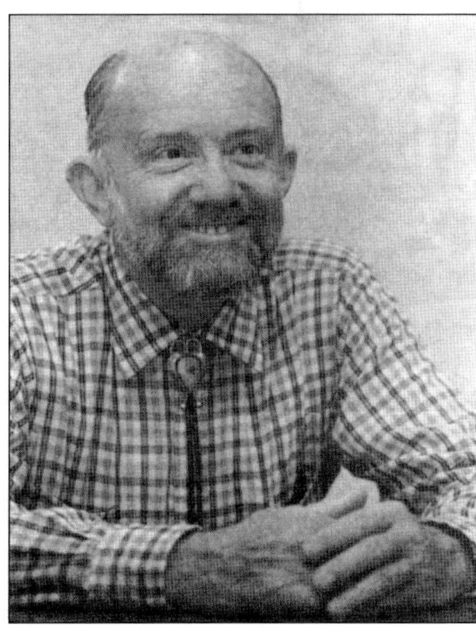

Jim Officer was another World War II veteran who went into radio announcing, first in Phoenix and then in Tucson at KTUC. In his words, he "defected to KCNA with Wayne Sanders in 1947," becoming the voice of sports while he was studying at the University of Arizona. Officer is still recognized as having been one of the best play-by-play sportscasters in the state. He joined the faculty of the University of Arizona in 1969 as a professor of anthropology. (Pauline Jackson.)

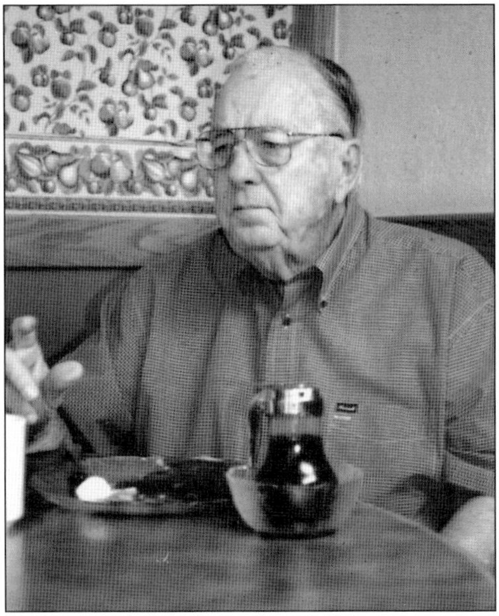

Tucson's fifth radio station was KTKT, which came on the air with 250 watts at 1490, at 7:00 p.m. on Tuesday, December 13, 1949. Veteran Tucson broadcaster George Wallace (shown here) came to town from Chicago with his father and brother, Tom Sr. and Tom Jr., to operate the new station. Radio star Tom Brenneman was an original partner, and radio-television star Art Linkletter invested in 1957. When the three Wallaces arrived to take over, they also went on the air as disc jockeys. The station had an opening-day library of 6,500 records. (Larry Schnebly.)

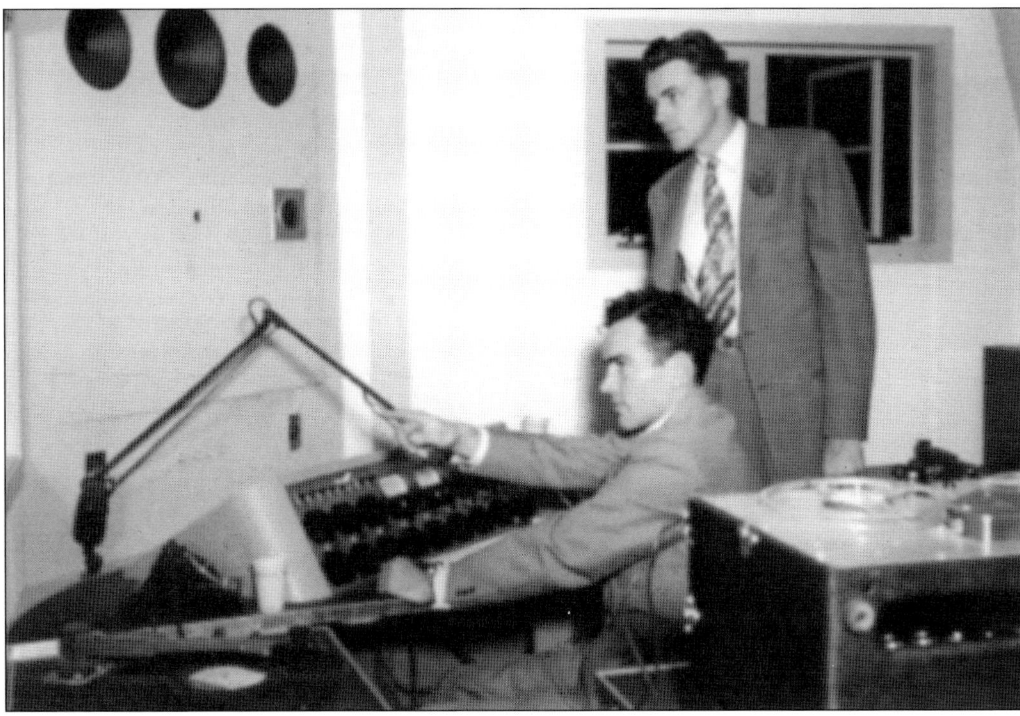

KTKT's first broadcast was an opening ceremony the night before regular programming began. Dignitaries, including Arizona governor Dan Garvey, called in, and congratulatory messages were received and read on the air. Tucson mayor E.T. Houston got lost trying to find the West Elm Street station, but he arrived in time to extend his good wishes. Before the Wallaces took over, Phillip Hurlbut was the manager-engineer, and his wife, Bernice, was the office manager. Here, Phil Hurlbut (standing) and announcer Larry Dietz are in the new control room directing the festivities. The station operated from these studios until it changed frequency to 990 and moved downtown in October 1957. (Ray Lindstrom.)

Chuck Blore was KTKT's afternoon personality in the early 1950s with *Let's Play Records*. He opened each show with a skit that ended with the name of the show leading into the theme song. Blore daily demonstrated his huge talent for radio. A talent scout heard Blore on KTKT, and he was hired by the legendary Gordon McLendon, who made him program director of his El Paso station. From there, Blore took his version of Top 40 to KFWB, channel 98, in Los Angeles. In a 2006 letter to coauthor C.J. Brown, Blore said, "I copied McLendon, and a zillion others copied me." Blore started a company that became one of the country's top producers of commercials. (CJB.)

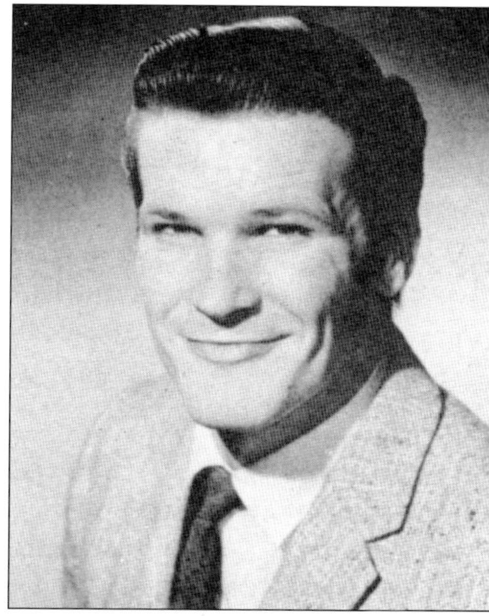

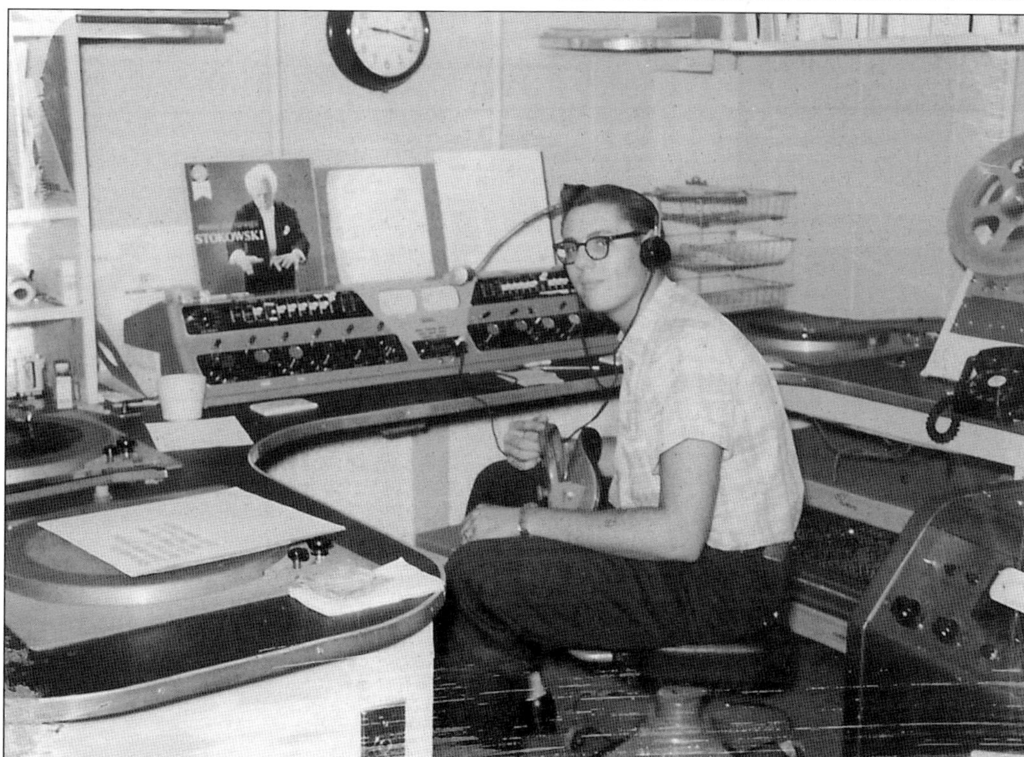

FM radio in Tucson did not exist until March 1954, when Tom Wallace put KTKT-FM on the air at 99.5. The format was classical music, and the music was supplied by the Wallaces' extensive record collection of Brahms, Beethoven, Bach, Mozart, and other great composers. There were specialty shows, too, like an hour of Gilbert and Sullivan music. The tape recorders seen here enabled KTKT-FM to play the new stereo recordings. On the air, stereo required two radio stations and KTKT AM/FM combined to do Tucson's first stereo demonstration programs. (RJ.)

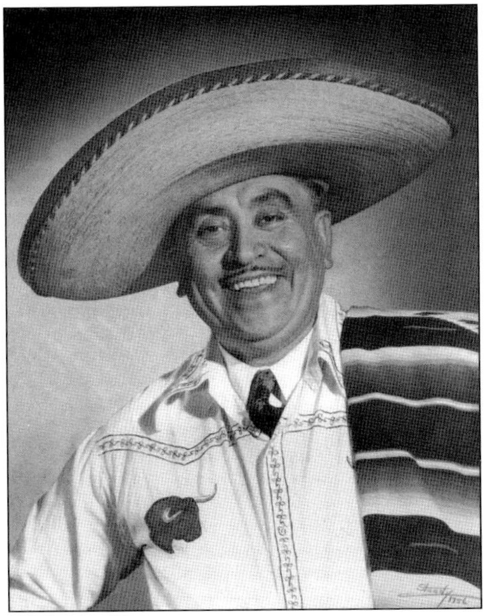

Spanish-language broadcasts have been a continuing presence in Tucson. The pioneer in this form of community-based programming was Jacinto Orozco, whose early-morning programs were a fixture from the 1940s. He set the standard for all who followed him. Born in Zacatecas, Mexico, Orozco came to Tucson in 1938 from the mining town of Jerome in northern Arizona. He worked at the Southern Pacific Railroad roundhouse during World War II while also doing his radio show on KVOA. His family says that older Tucsonans stop to tell them stories of Don Jacinto and what he meant to them. Tucson's Mexican community was brought closer together by his broadcasts. (Bob Orozco.)

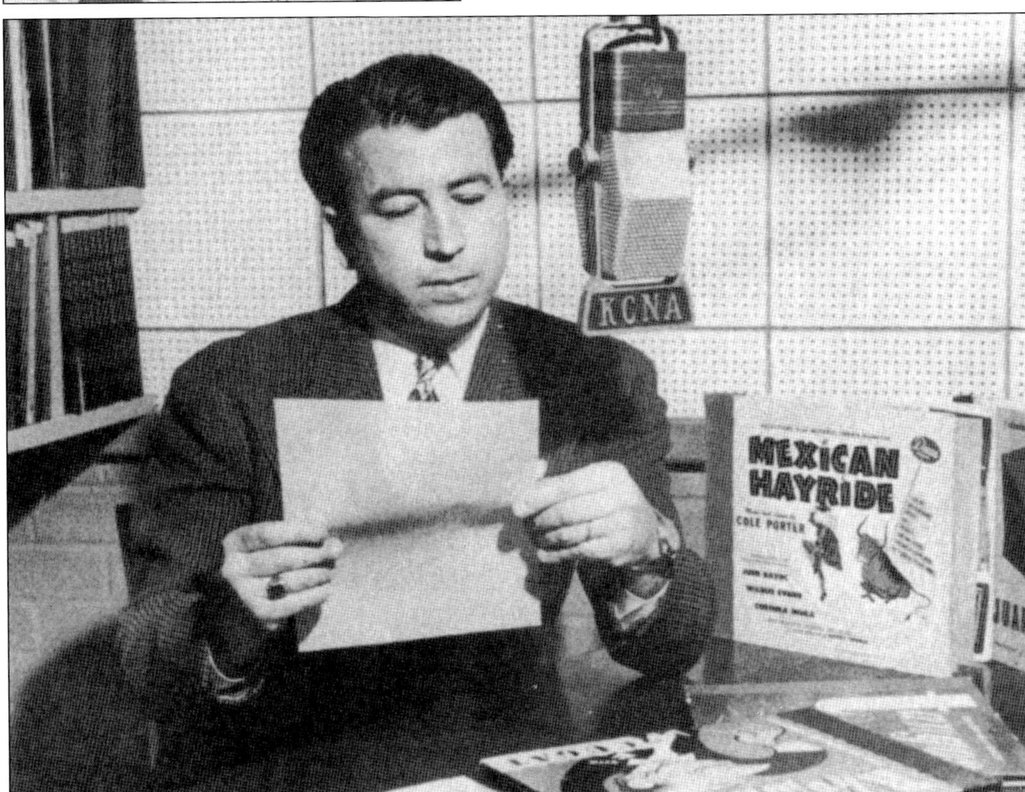

The other Tucson radio stations hired their own Spanish-speaking announcers to do programs. Guillermo Sosa "Bill" Lopez (seen here) was on KCNA daily in the 1940s. C.B. Sedillo Jr. was on KOPO; Tony Villegas, KTUC; and Manny Ornelas, KTKT and then KCNA in the evenings. Tucson's first station to broadcast exclusively in Spanish came in 1953, when easterner Pete Trowbridge opened daytime-only KEVT 690 along with a radio repair and sales store. (CJB.)

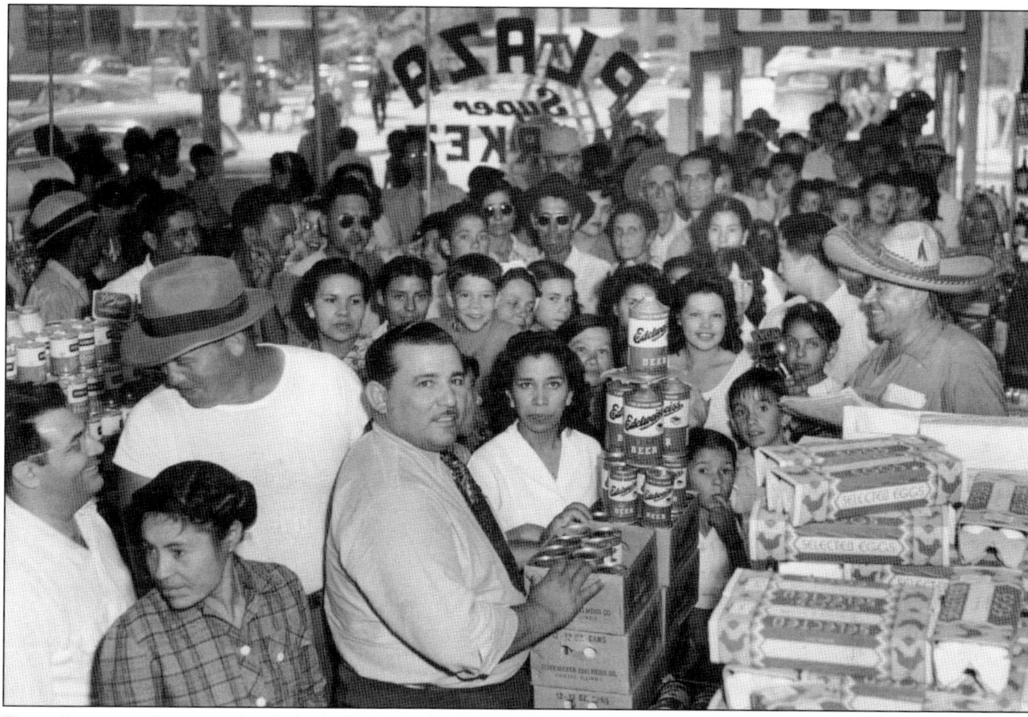

Don Jacinto Orozco (right) did many broadcasts from the community, like this one at Plaza Market in 1946 on KVOA. Listeners turned out in large numbers to greet him, always wearing his trademark sombrero. The boy with the bangs in front of Orozco is his son Bob. Brothers Bob and John were the radio program producers and became Jacinto's television program producer-directors at channel 4. When KVOA sold 1290, Orozco moved his radio program to KTAN (formerly KCNA on 580), whose 5-kilowatt signal was carried to many new listeners throughout southern Arizona. (Bob Orozco and Joe Howell.)

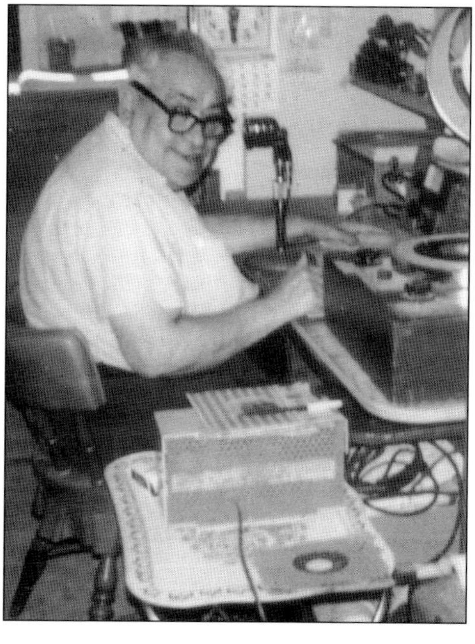

The Jacinto Orozco program was broadcast from this studio in his home on South Eighth Avenue, in Tucson's Barrio Libre. KVOA's engineers set him up with one turntable, requiring him to change records while he talked on-air. His measured and somewhat hoarse voice referring to his listeners as "amigos mios," along with his impressive list of sponsors and constant supply of fan mail, made him a legacy to be admired in Tucson broadcast history. After KTAN's sale, Orozco moved his program to KEVT, where he stayed until he retired. (Bob Orozco.)

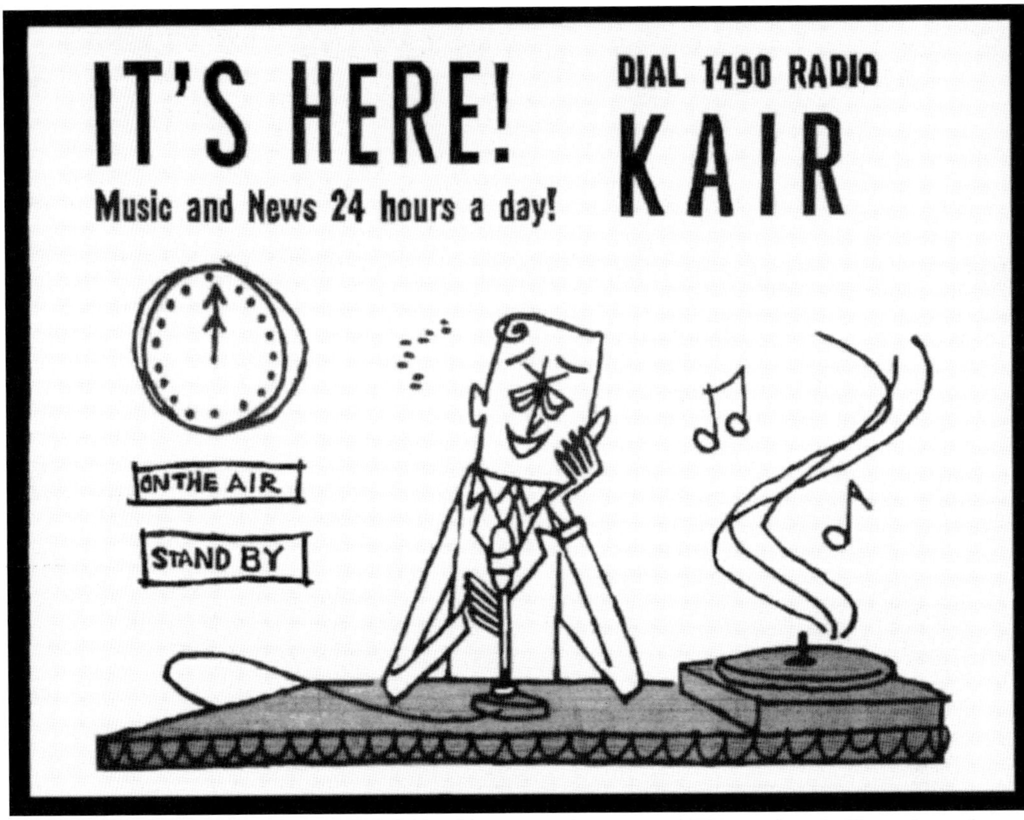

KAIR had a unique beginning. KTKT had been on the air on 1490 as a local all-music and news station since 1949. When the Wallace family moved their station to 990 with the big 10-kilowatt signal in January 1957, new owners jumped onto 1490 and continued the same format so that listeners would not recognize the difference. KAIR was housed in the former KVOA studios at 48 East Broadway. Dan Gates was one of the first KAIR announcers. They were required by the owners to play one Frank Sinatra song every hour. (CJB.)

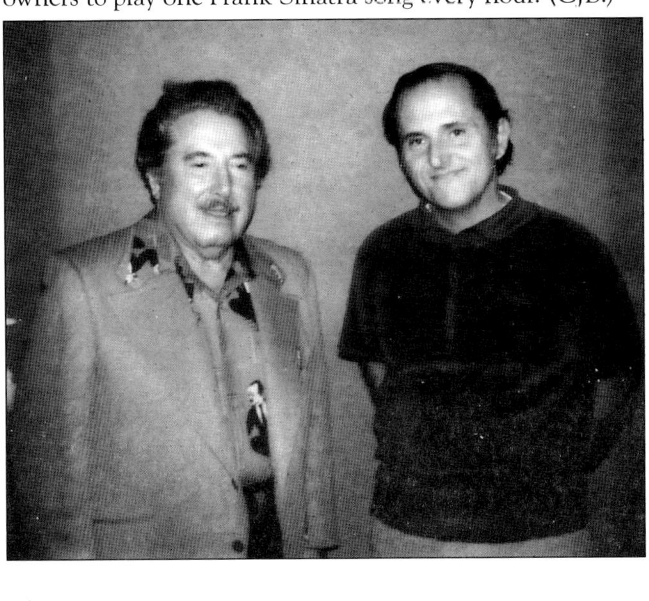

Harold Peary starred on NBC radio in *The Great Gildersleeve*, a comedy series about a water commissioner, his family, and friends. The character of Throckmorton P. Gildersleeve was first heard as a visitor to Fibber McGee and Molly on their radio show. The role was spun off, not only on network radio, but in several movies built around the character. Peary's infectious laugh was his trademark. In this 1970s photograph, he is being interviewed by coauthor Russ Jackson (right). (Susan Jackson.)

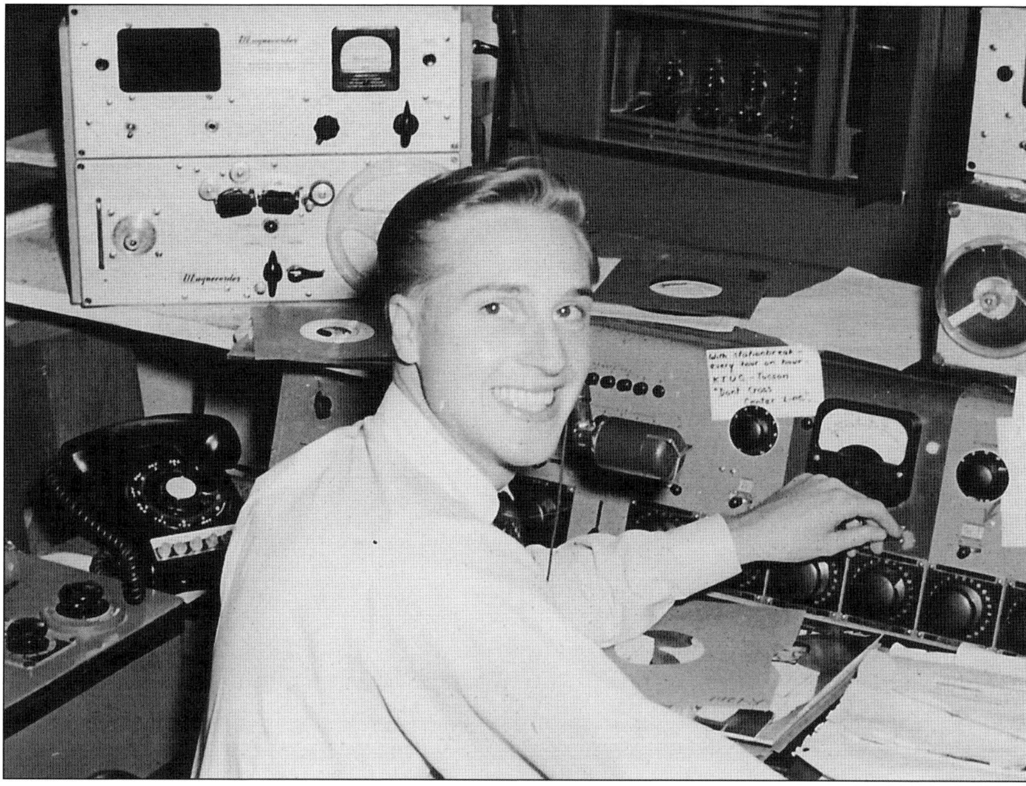

Chris Borden's career in broadcasting started with KVOA, on a program with his UA fraternity brothers. KTUC was the first station to broadcast 24 hours a day, and Borden went there in 1956. The show, from 12:00 a.m. to 3:00 a.m., had plenty of listeners, he says. "I did contests, had late night guests, had musicians play the studio's beautiful grand piano, set up an outside, lighted 'booth' and cars honked as they drove by during the warm summer nights. And, I made . . . get this . . . about $200 a week in advertising revenue! The sponsors loved it. This photo of me is a hoot!" (Chris Borden.)

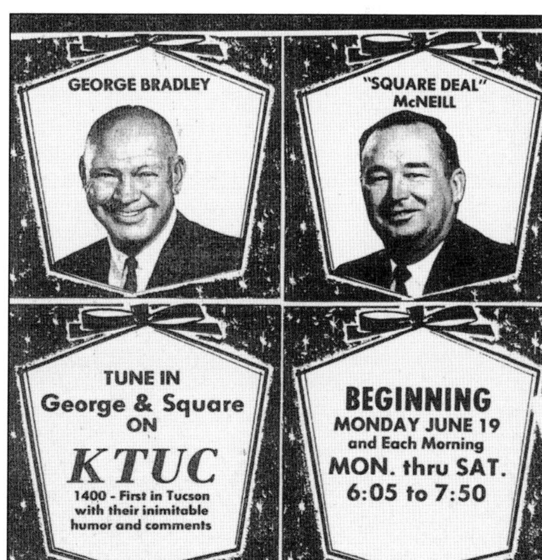

George Bradley and "Square Deal" McNeil were heard on KTUC and did gigs for KTAN. In addition, Bradley opened KFIF. Chris Borden remembers his first days at KTUC: "I went on the air at 6:00 a.m., doing just engineering work and spinning music from those big 16-inch transcriptions for George and Square. Then [station manager] Lee Little came in and did *45 Minutes from Broadway*." KTUC had *Mutual's Game of the Day*, daytime Major League Baseball games broadcast live with hall-of-fame pitcher Dizzy Dean doing play-by-play, in addition to its local sports broadcasts. (CJB.)

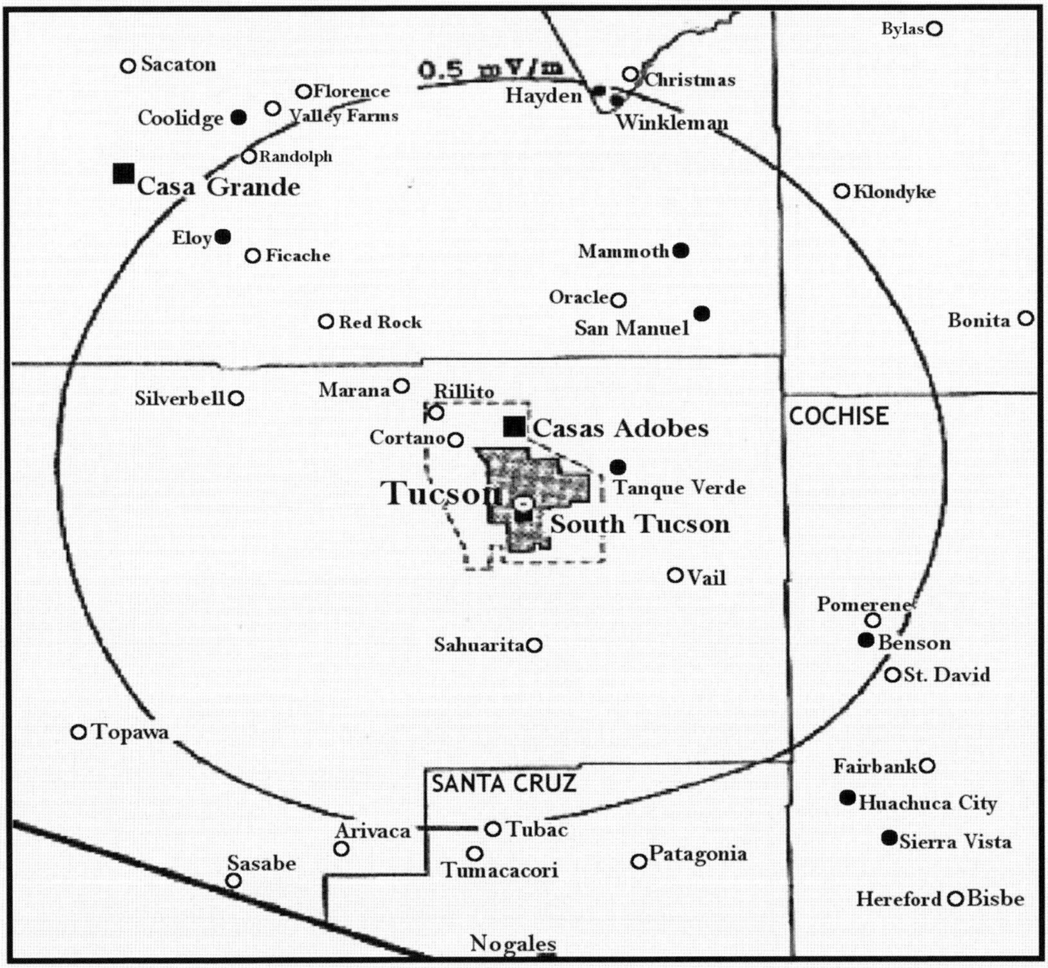

All the radio stations in the country that had been limited to 250 watts of power were allowed to apply for increases to 1 kilowatt. KTUC on 1400, KOLD on 1450, and KAIR on 1490 received this increase, which extended their Tucson market coverage areas, as this KTUC map shows. On KVOA, which already had 1 kilowatt, program director Larry Schnebly was hosting a midday show, Bill Rudd came from WDGY Minneapolis to do mornings, and Clare Coffman arrived from KVSF in Santa Fe to do afternoons. In 1956, the KVOA radio-television stations were sold to Clinton D. McKinnon, and in 1958, the station was sold and the call letters were changed to KCUB. (CJB.)

Three

Top 40 Rules
Playing Rock 'n' Roll All the Time

Top 40 radio was first heard in the early 1950s, when Todd Storz and Gordon McLendon separately debuted the format on stations they owned. Both men possessed savvy broadcast programming and marketing skills. Using these tools, combined with a format playing only the most popular music in America, gave both men great successes in radio by the mid-1950s. At the same time, America's youngsters were getting excited about a new form of music. First heard in the early 1950s, it was a combination of elements heard in country-western and rhythm-and-blues music. This new music was given the name "rock 'n' roll" by Cleveland, Ohio, DJ Alan Freed. With the dramatic rise to fame of Elvis Presley in the mid-1950s, rock 'n' roll gained a powerful voice and face to help spread its popularity. As this new music spread, record sales soared, fueled by a growing source of disposable income from American teens. These increased sales were reflected in charts published in the music industry's leading trade magazines, *Billboard* and *Cashbox*. As a result of these large record sales and chart successes, more and more of the music heard on radio was rock music. Top 40 music became mostly rock 'n' roll music. Teens flocking to these Top 40 stations attracted the attention of sponsors eager to get their piece of the baby boomers' spending.

The growth of Top 40 radio was also a result of the television revolution in the early to middle 1950s. During this same time, Americans were deserting the family radio to embrace the pictures and sounds available from television. By the mid-1950s, networks and sponsors were quickly moving to television. This network cutback created big programming voids at many radio stations. Owners and operators soon discovered that one of the fastest and cheapest options to fill programming needs was to have a disc jockey play records. Such a program lent itself to any kind of music. But it did not take long for station operators to realize that rock music, with its legion of teen fans, offered the greatest listener response and sponsor excitement.

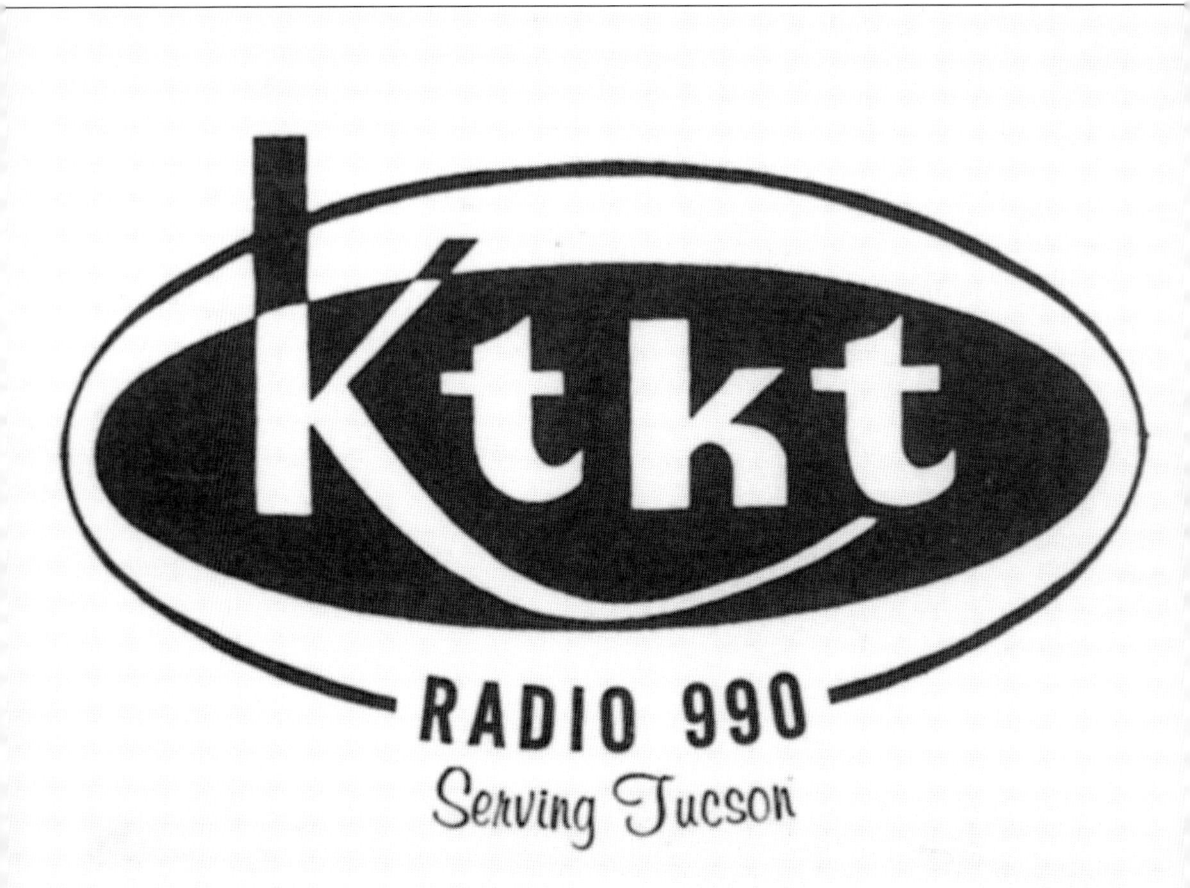

Eventually, Tucson radio followed the national trend and jumped on the rock 'n' roll train. A few Top 40 radio programs were first heard in Tucson during afternoon and evening hours as early as 1955. The first Tucson station to play Top 40 full time was KTKT. Top 40 music and an exciting new format called "Color Radio" made its Tucson debut in May 1957. Not only was the music exciting, but the other program elements, such as news, weather, and even the advertisements, were presented in a fast-paced, exciting manner designed to keep the listener tuned in just to hear what would be played next. This exciting new format took Tucson by storm. As a result, KTKT would dominate the Tucson ratings for about the next 25 years. The mastermind behind the KTKT Color Radio format was Tucsonan Frank Kalil. Prior to KTKT, Kalil worked for Top 40 pioneer Gordon McLendon in Texas under Chuck Blore. This gave Kalil the opportunity to learn the latest programming techniques of the era. During the late 1950s, a couple of other Tucson stations, KAIR and KCUB, tried to take on KTKT, but soon gave up. During the 1960s, KFIF and, later, KIKX, made serious efforts to dethrone KTKT, but they also failed. During the 1970s, KTKT faced challenges from a rejuvenated KIKX and KHYT on the AM side, while FM rock came to Tucson in 1970 with KWFM. In July 1977, KRQQ became Tucson's first FM Top 40 station. By the early 1980s, further FM penetration, along with format and audience fragmentation, caused KTKT to lose its ratings throne. (Ray Lindstrom.)

Before there was Color Radio KTKT, and before there was a rock and roll dominated Top 40, from the 1940s through the 1950s every night on 48 radio stations in 11 western states there was Lucky Lager Dance Time. Between the program's opening theme, Benny Goodman's "Let's Dance," until the closing theme, "Dream" by the Pied Pipers at Midnight, the top-selling records of the day were played from a master list prepared by Bill Gavin in San Francisco that was played on all stations. In Tucson, the first host on KCNA was Wally Lindberg. When the program moved to KVOA, its host, new announcer Larry Schnebly, remembers one night at Christmastime he came home to find a gift case of the sponsor's product on the doorstep. (CJB.)

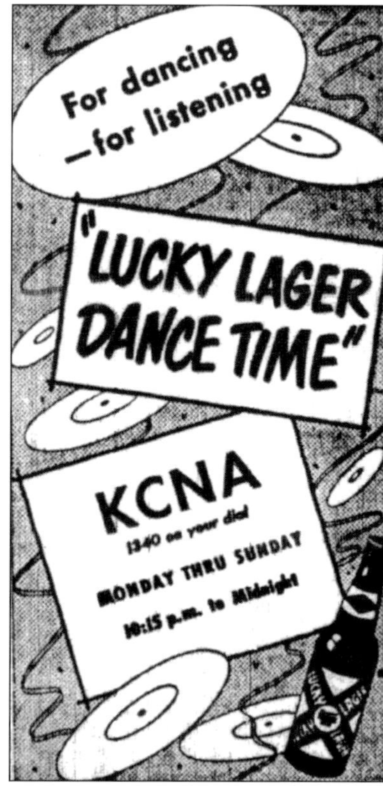

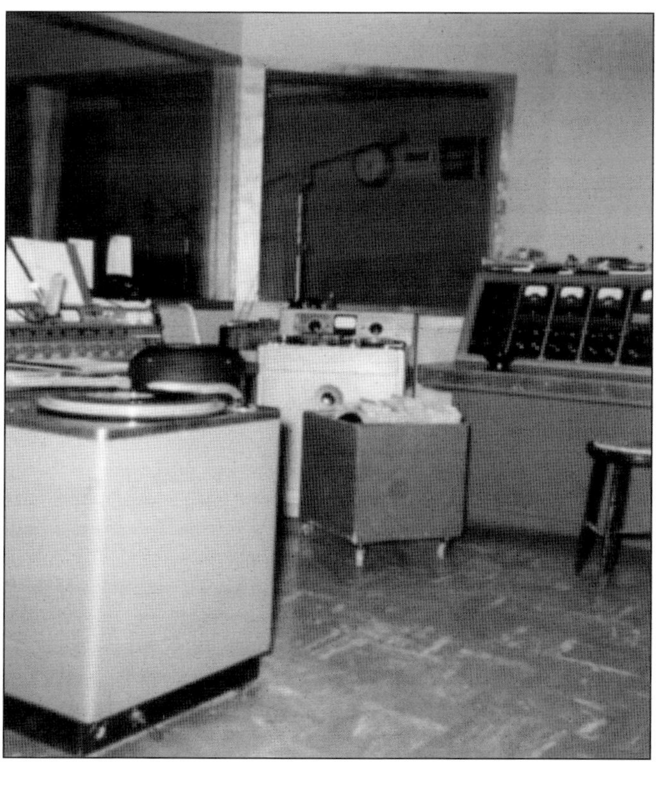

When Lucky Lager expanded *Dance Time* to the nine-station ABS network, including KVOA, the two-and-a-half-hour show originated in this KTAR master control room in Phoenix, with its big network switcher. Neil Sargent, the statewide voice of the program, remembers that a friend told him years later that he had proposed to his future wife while they listened to Sargent do *Dance Time* on Flagstaff affiliate KCLS. After the mid-1950s, most of the top-selling songs were rock tunes, and parents began to object to a beer company sponsoring a show that was now playing music that appealed to teenagers. Despite messages that advised minors not to buy its product, the program was canceled in 1960. Competition from Top 40 stations was growing, too. (RJ.)

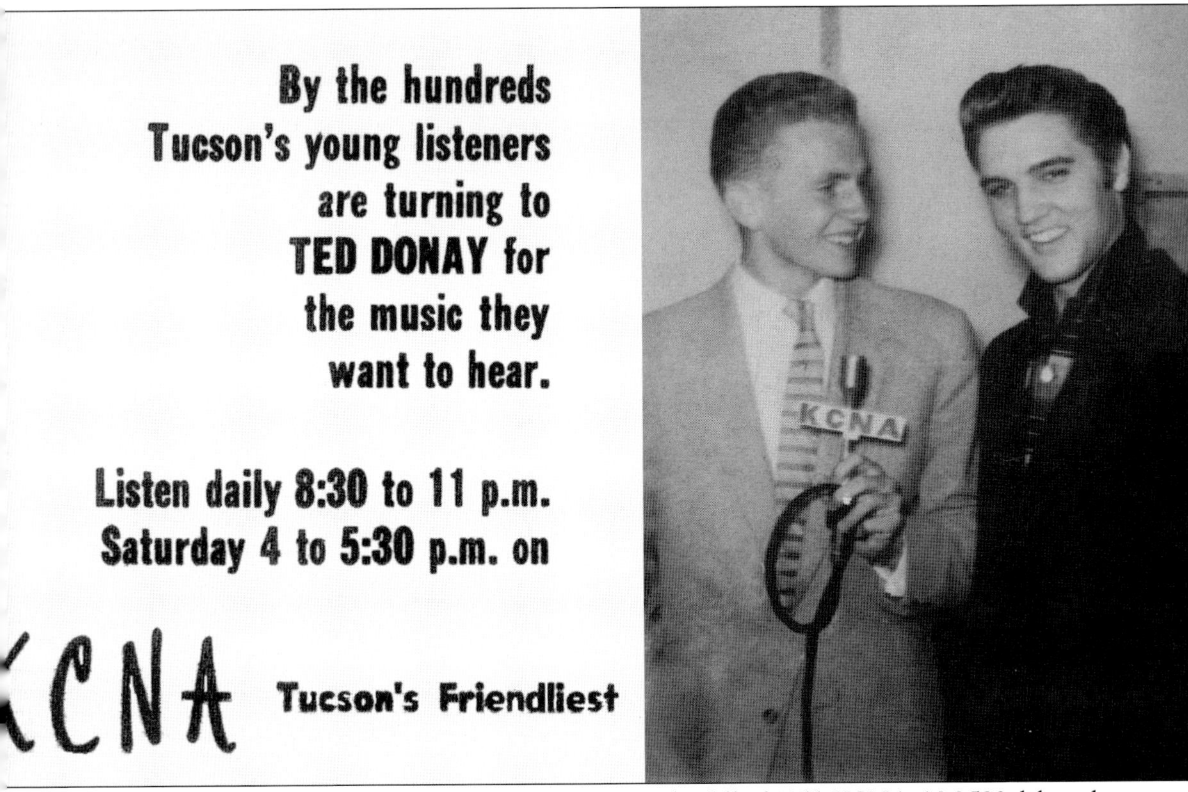

Top 40 radio may have first been heard in Tucson in the fall of 1955. KCNA AM 580 debuted an afternoon rock 'n' roll music program following the 5:00 p.m. news. Hosted by Russ Wonnacott, this program, intended for the teenage listener, ran until 6:00 p.m. weekdays and was called *Rock and Roll with Russ*. About the same time, KCNA added another rock program in the evening: Ted Donay, a UA student, hosted *The College Dance Hall* from 8:00 p.m. to 11:00 p.m. on weeknights. According to Tucson radio veteran Dan Gates, Donay became popular with the young listeners by playing Elvis Presley records long before they were heard elsewhere. KCNA and Ted Donay both received a big boost when Elvis appeared at the Tucson Rodeo Grounds on June 10, 1956. Donay interviewed Presley on that day, and the taped interview was later aired on his KCNA program. Donay worked hard to make his show a success, and he earned praise and accolades in local newspapers from KCNA owner Harry Chambers and sponsors. Donay left Tucson in early 1957. In late 1958, KCNA went to a full-time easy listening format. (Shay Oliver.)

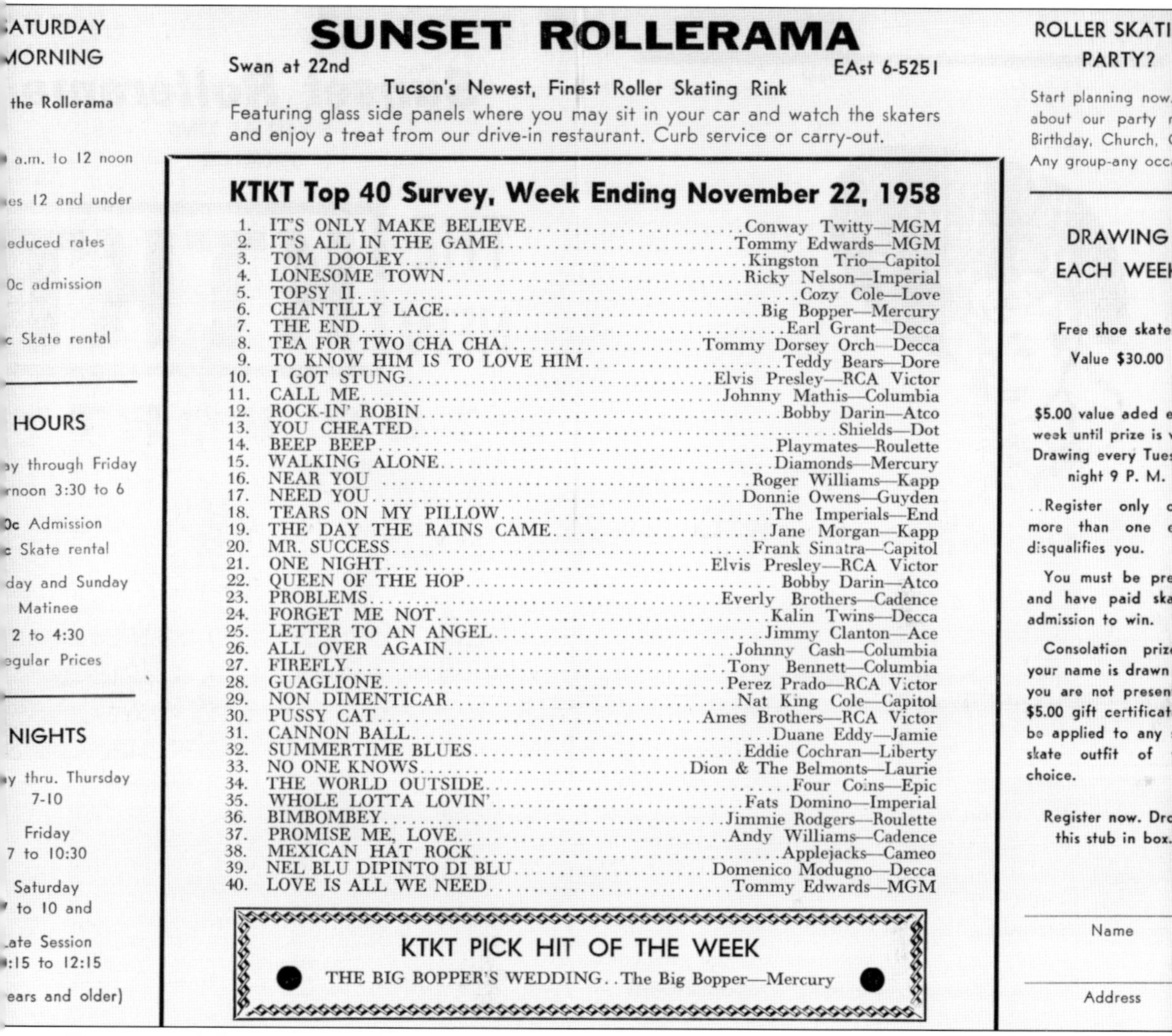

SATURDAY MORNING

the Rollerama

a.m. to 12 noon

es 12 and under

educed rates

0c admission

c Skate rental

HOURS

ay through Friday

rnoon 3:30 to 6

0c Admission

c Skate rental

day and Sunday Matinee

2 to 4:30

egular Prices

NIGHTS

ay thru. Thursday 7-10

Friday 7 to 10:30

Saturday to 10 and

ate Session :15 to 12:15

ears and older)

SUNSET ROLLERAMA

Swan at 22nd — EAst 6-5251

Tucson's Newest, Finest Roller Skating Rink

Featuring glass side panels where you may sit in your car and watch the skaters and enjoy a treat from our drive-in restaurant. Curb service or carry-out.

KTKT Top 40 Survey, Week Ending November 22, 1958

	Title	Artist—Label
1.	IT'S ONLY MAKE BELIEVE	Conway Twitty—MGM
2.	IT'S ALL IN THE GAME	Tommy Edwards—MGM
3.	TOM DOOLEY	Kingston Trio—Capitol
4.	LONESOME TOWN	Ricky Nelson—Imperial
5.	TOPSY II	Cozy Cole—Love
6.	CHANTILLY LACE	Big Bopper—Mercury
7.	THE END	Earl Grant—Decca
8.	TEA FOR TWO CHA CHA	Tommy Dorsey Orch—Decca
9.	TO KNOW HIM IS TO LOVE HIM	Teddy Bears—Dore
10.	I GOT STUNG	Elvis Presley—RCA Victor
11.	CALL ME	Johnny Mathis—Columbia
12.	ROCK-IN' ROBIN	Bobby Darin—Atco
13.	YOU CHEATED	Shields—Dot
14.	BEEP BEEP	Playmates—Roulette
15.	WALKING ALONE	Diamonds—Mercury
16.	NEAR YOU	Roger Williams—Kapp
17.	NEED YOU	Donnie Owens—Guyden
18.	TEARS ON MY PILLOW	The Imperials—End
19.	THE DAY THE RAINS CAME	Jane Morgan—Kapp
20.	MR. SUCCESS	Frank Sinatra—Capitol
21.	ONE NIGHT	Elvis Presley—RCA Victor
22.	QUEEN OF THE HOP	Bobby Darin—Atco
23.	PROBLEMS	Everly Brothers—Cadence
24.	FORGET ME NOT	Kalin Twins—Decca
25.	LETTER TO AN ANGEL	Jimmy Clanton—Ace
26.	ALL OVER AGAIN	Johnny Cash—Columbia
27.	FIREFLY	Tony Bennett—Columbia
28.	GUAGLIONE	Perez Prado—RCA Victor
29.	NON DIMENTICAR	Nat King Cole—Capitol
30.	PUSSY CAT	Ames Brothers—RCA Victor
31.	CANNON BALL	Duane Eddy—Jamie
32.	SUMMERTIME BLUES	Eddie Cochran—Liberty
33.	NO ONE KNOWS	Dion & The Belmonts—Laurie
34.	THE WORLD OUTSIDE	Four Coins—Epic
35.	WHOLE LOTTA LOVIN'	Fats Domino—Imperial
36.	BIMBOMBEY	Jimmie Rodgers—Roulette
37.	PROMISE ME, LOVE	Andy Williams—Cadence
38.	MEXICAN HAT ROCK	Applejacks—Cameo
39.	NEL BLU DIPINTO DI BLU	Domenico Modugno—Decca
40.	LOVE IS ALL WE NEED	Tommy Edwards—MGM

KTKT PICK HIT OF THE WEEK

THE BIG BOPPER'S WEDDING. . The Big Bopper—Mercury

ROLLER SKATI PARTY?

Start planning now. about our party r Birthday, Church, C Any group-any occ

DRAWING EACH WEEK

Free shoe skates Value $30.00

$5.00 value aded e week until prize is w Drawing every Tues night 9 P. M.

Register only o more than one e disqualifies you.

You must be pre and have paid ska admission to win.

Consolation prize your name is drawn you are not present $5.00 gift certificate be applied to any s skate outfit of choice.

Register now. Dro this stub in box.

Name

Address

Top 40 as a full-time format arrived in Tucson in 1957. KTKT founder and owner Tom Wallace made several changes that left a big mark on Tucson radio history. First off was the frequency shift from AM 1490 to 990 to allow for a stronger, 10,000-watt signal. The drawback was that the station was granted a daytime-only license. The frequency shift occurred on January 5, 1957. KTKT AM and FM moved to new studios and offices downtown in October 1957. The FM transmitter and antenna would also move downtown. The AM transmitter and antenna had already moved to West Grant Road when the switch was made to 990. To help pay for these changes, Wallace sold a 49-percent stake in KTKT to Art Linkletter in 1957. Wallace had known Linkletter since the early 1940s while working in the advertising and radio industries in Chicago. One other change had a lasting impact. In the spring of 1957, Wallace hired Frank Kalil to program KTKT. Kalil launched Color Radio, immediately making KTKT the market leader. The 1958 chart shown here offers an example of the music played on Color Radio. KTKT soon became the 800-pound gorilla of Tucson radio, with Frank Kalil serving as King Kong. (Frank Kalil.)

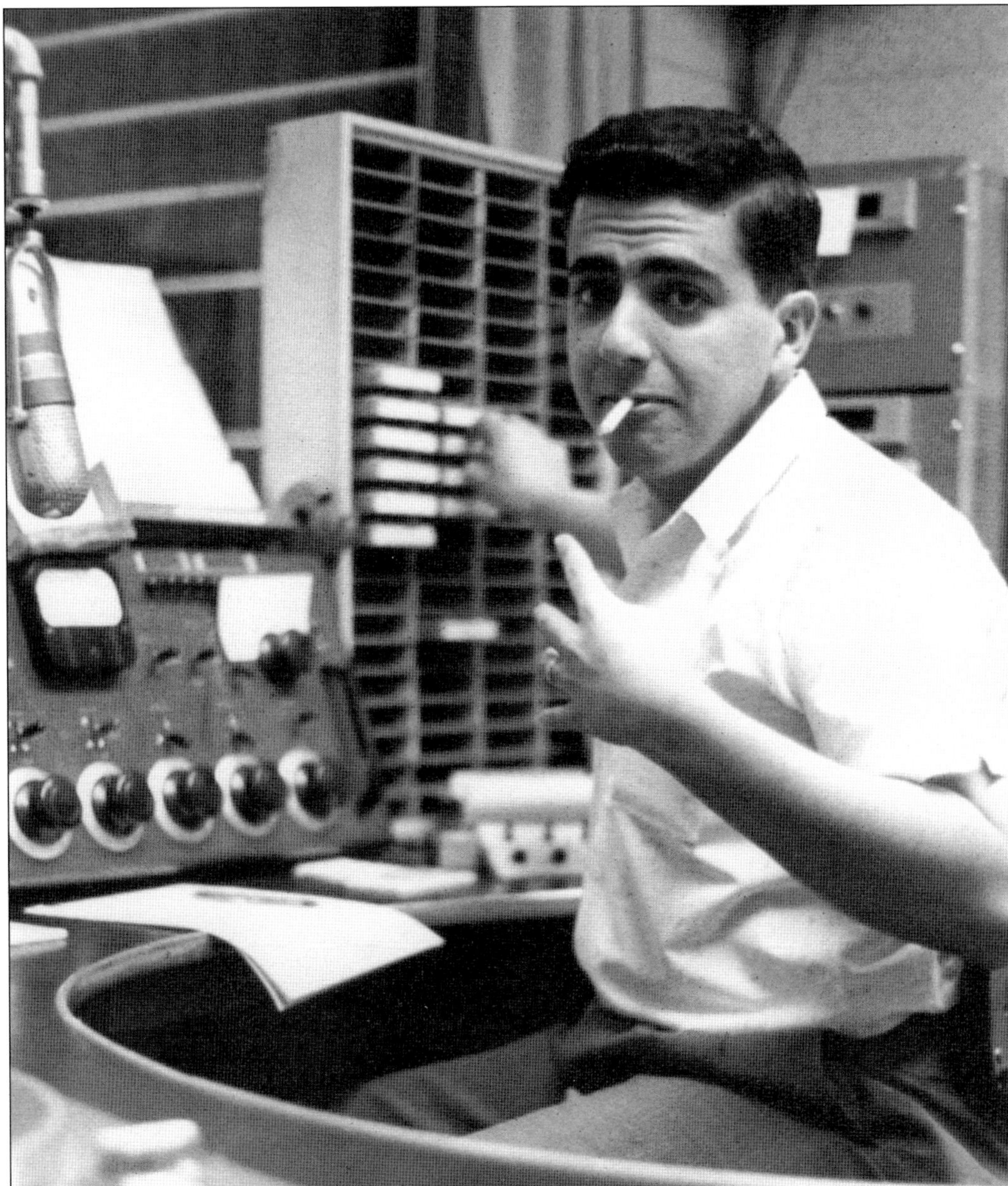

Frank Kalil met Chuck Blore at age 15 after winning a tryout to host a weekly program for teens on KTKT. Kalil started hanging out at the station all week to learn from Blore, who also helped Kalil prepare for his Saturday teen program. After a couple of years, Blore was recruited by Gordon McLendon, becoming PD at KELP in El Paso, Texas, where he put a polished Color Radio format in place. He soon hired Kalil, who soon got ratings of more than 50 percent, while Blore captured ratings of 72 percent. National industry notice of these ratings prompted KTKT owner Tom Wallace to fly to El Paso. He hired Frank Kalil to program KTKT. Now in Tucson, Kalil used everything he had learned from Blore and the Gordon McLendon group and added his own drive for excellence when he launched Color Radio at KTKT in May 1957. The results of his efforts allowed KTKT to dominate Tucson radio for more than 20 years. As Kalil said, "Color Radio blew the doors off everything in Tucson." (Frank Kalil.)

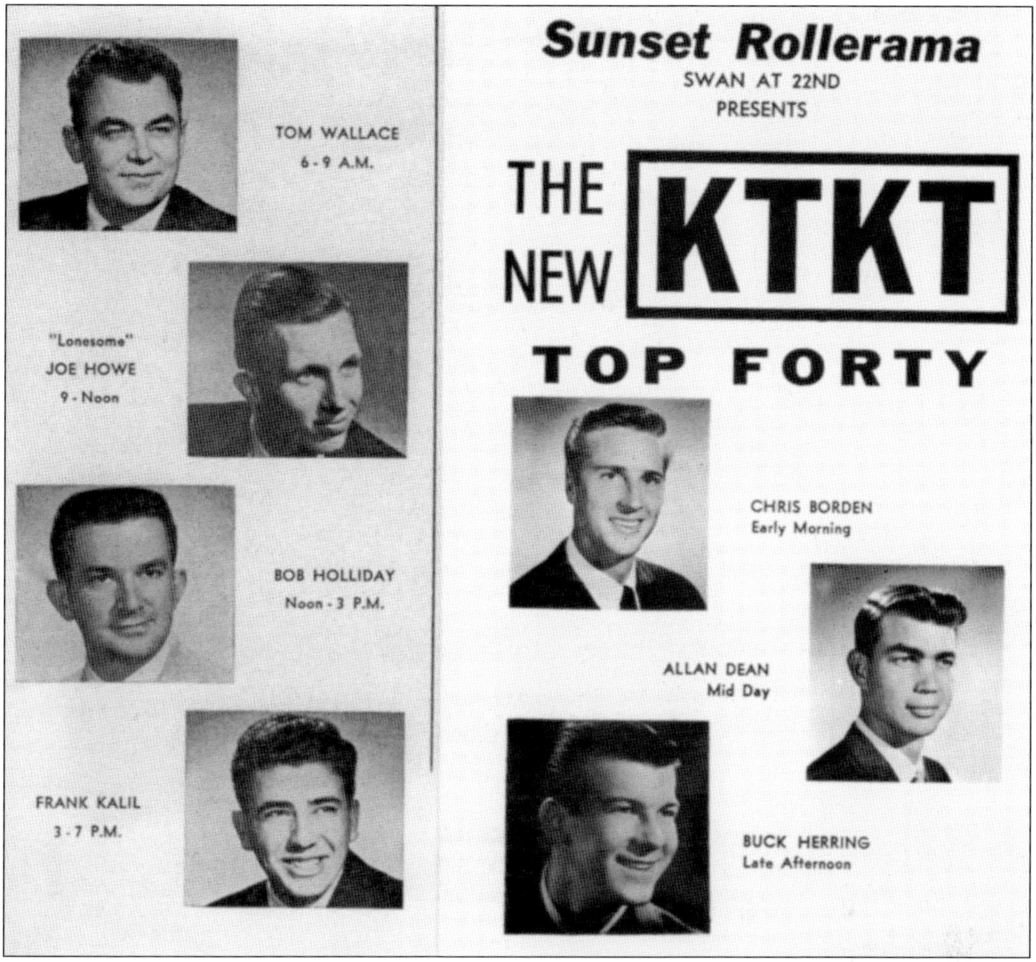

In the above image, the KTKT lineup on the left is from a 1957 brochure. The lineup on the right is from a November 1958 KTKT chart cover. Tom Wallace is the owner's son and KTKT chief engineer. Joe Howe went on to become a Maricopa County Superior Court judge. Bob Holiday had another tour of duty at KTKT during the late 1960s. Kalil was already getting top ratings in his time slot. On the 1958 lineup is morning man Chris Borden, whose show featured a lot of humor. Borden also operated a school for DJs in Tucson. After leaving Tucson, he had great success at KEWB in San Francisco. Buck Herring moved into Frank Kalil's time slot while Kalil was away working for Sherwood Gordon. Kalil returned to KTKT in the spring of 1959. After KTKT, Buck had further success in California. The photograph at right shows the on-air studio in 1958. (Above, Frank Kalil; right, RJ.)

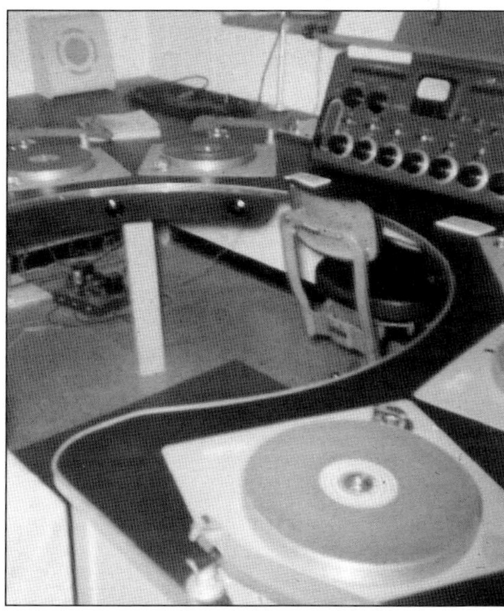

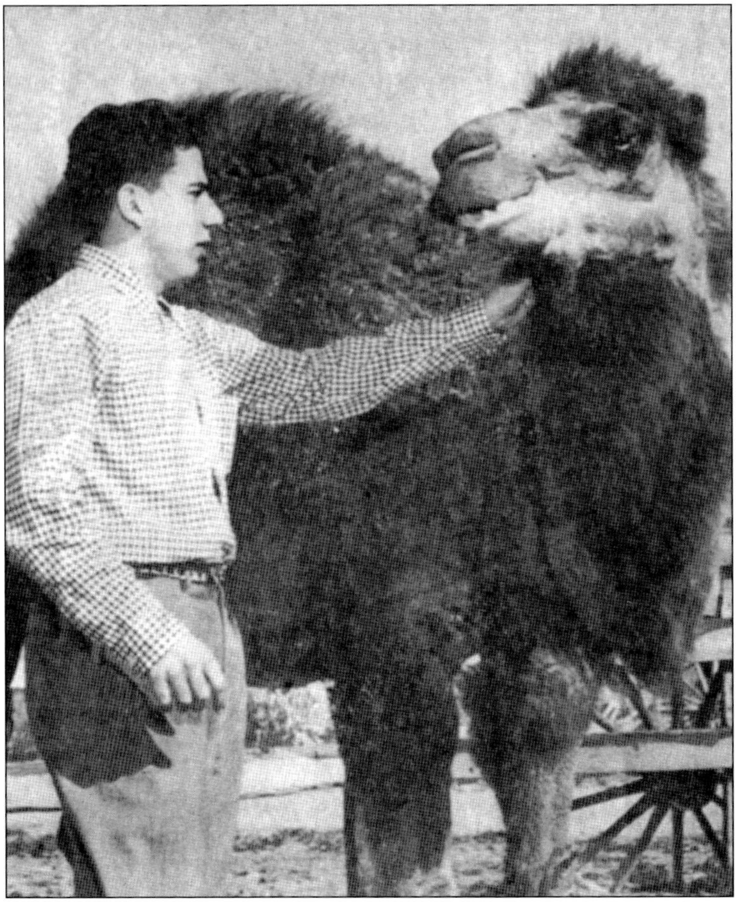

Frank Kalil learned how important promotion can be while working in the McLendon group. Chuck Blore taught Kalil the nuts and bolts of successful McLendon-style promotion. A few months after the launch of Color Radio, Kalil purchased a camel from a circus in El Paso that was facing bankruptcy. He named the camel Lily Kay Kalil, and she soon became a star. Lily Kay would appear with Frank at station promotions and at remote broadcasts. Kalil, who is Lebanese, would also appear in Tucson's annual Rodeo Day Parade in full Arab dress with Lily Kay. Tucson loved Frank, Lily Kay, and KTKT. (Frank Kalil.)

CITY: TUCSON, ARIZ. **City Zone** MONTHS: JULY THRU SEPTEMBER, 1957

SHARE OF *RADIO* AUDIENCE

TIME	RADIO SETS-IN-USE	KAIR	KCNA (ABC)	KEVT	KOLD (CBS)	KTKT	KTUC (MBS)	KVOA (NBC)					OTHER AM & FM	SAMPLE SIZE
MONDAY THRU FRIDAY 7:00 A.M.-12:00 NOON	13.5	25.1	5.6	6.3	8.0	21.9	17.3	11.0					4.9	9,377
MONDAY THRU FRIDAY 12:00 NOON- 6:00 P.M.	11.3	26.4	4.3	5.2	7.0	27.9	15.9	9.5					3.8	11,134

Pictured here are ratings results from the summer of 1957, right after the launch of Color Radio. Kalil is already number one after a month. A short time later, KTKT ratings would lead in all time periods. (Frank Kalil.)

The photograph at right shows Frank Kalil taking part in a very dangerous stunt on February 19, 1961. He entered a bullfight at the Plaza de Toros in Nogales, Sonora, Mexico, just across the Arizona border. Kalil said that when the bull's horns scraped his leg at one point, it felt like his leg was being amputated. The below photograph shows two DJ lineups. The one on the left is from March 1959, and the one on the right is from January 1960. Note that Chris Borden is now PD. Johnny Hyde, previously at KCNA, took over Kalil's time period. Bob Sinclair moved on to KTSA in San Antonio, Texas, after KTKT. Russ Jackson, previously at KCNA and KVOA, would later work at KAIR, KMOP, and KTAN. Alan Carroll voiced the famous KTKT news closer that would be heard in Tucson for 10 years: "If you don't want it heard on KTKT news, don't let it happen" (Right and below right, Frank Kalil; below left, RJ.)

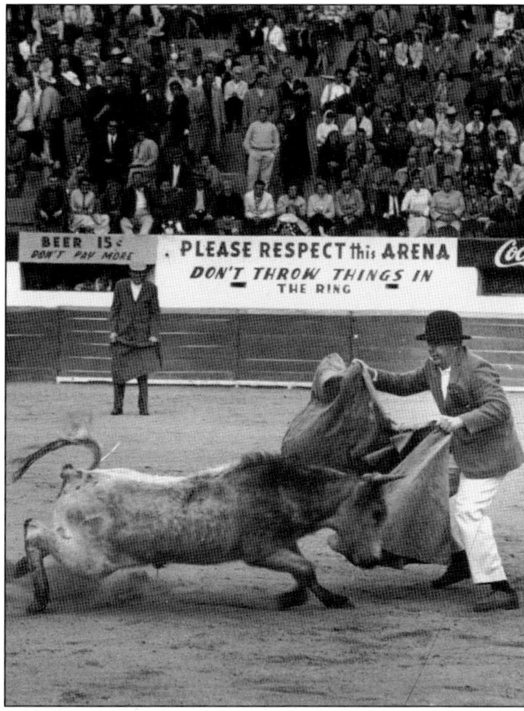

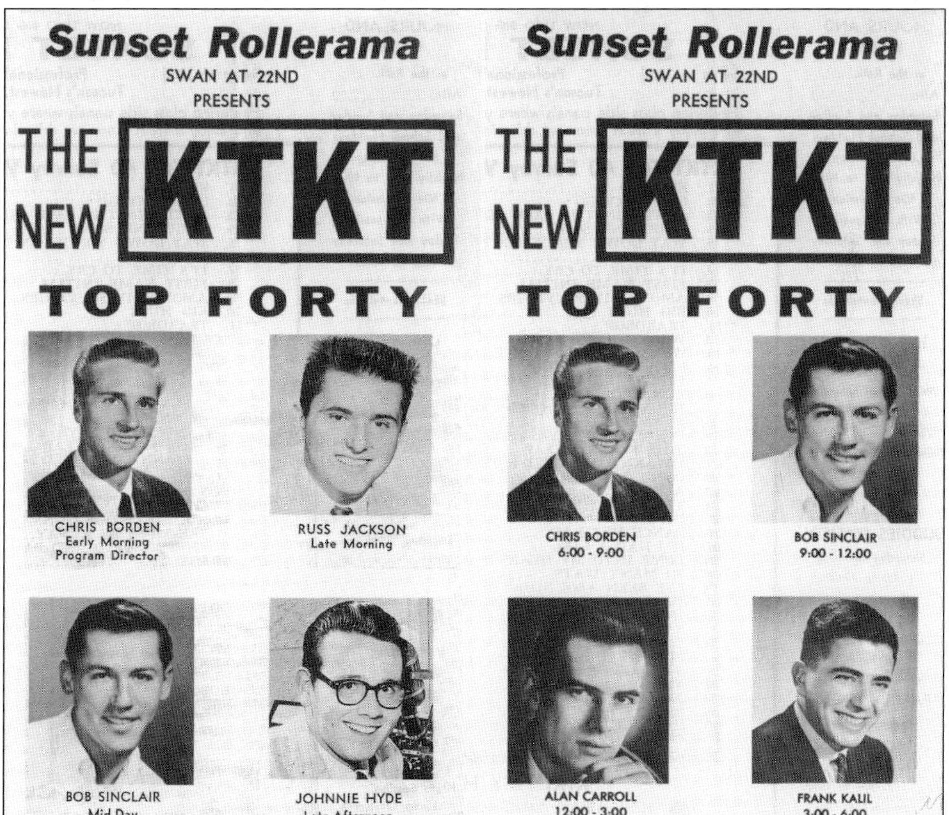

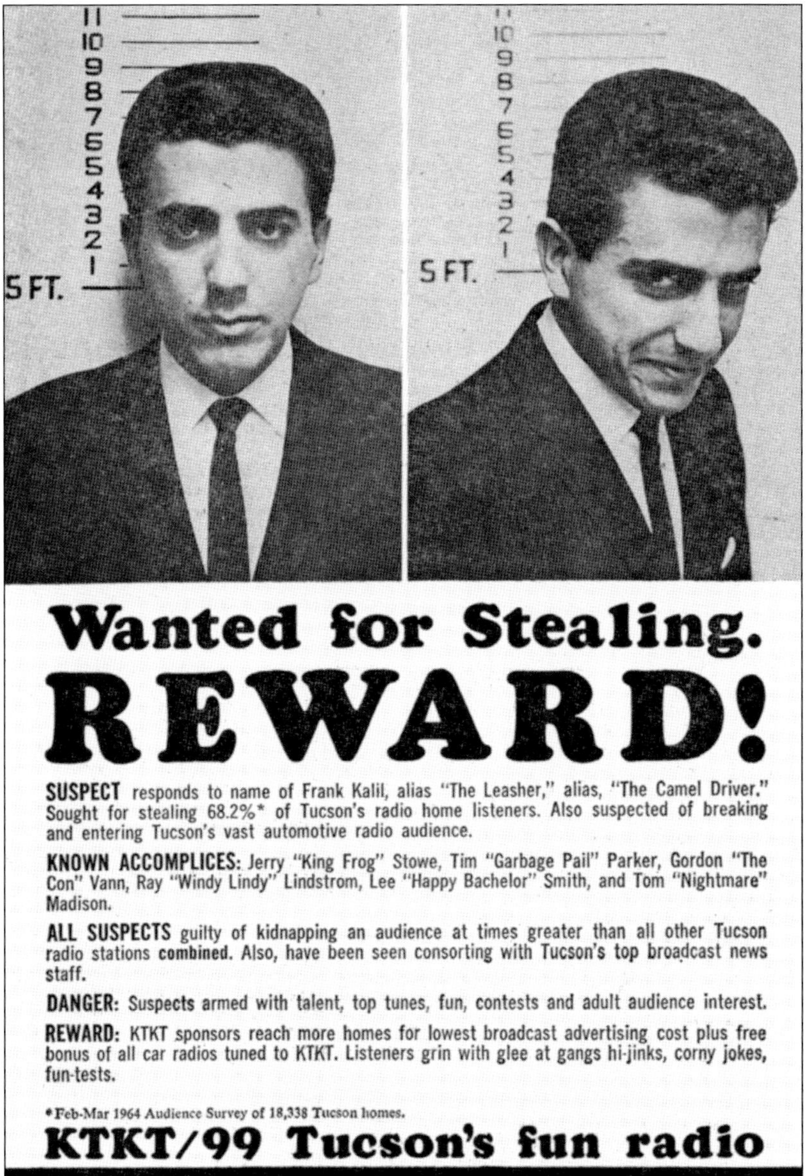

KTKT tightened its grip on Tucson audiences after being granted 24-hour authority during April 1960. Next, LeLand Bisbee purchased KTKT in October 1960. Phil Richardson, previously at KRUX in Phoenix, also owned by Bisbee, became general manager. The combination of Richardson's business skills and Frank Kalil's programming genius and drive for perfection created a perfect storm for success at KTKT. This advertisement touts Frank Kalil's recent ratings of more than 60 percent in early 1964. KTKT's consistent overall high ratings attracted sponsors eager to reach its large audience. Except for two newscasts totaling about five minutes, including a pair of 60-second advertisements, the format was one record followed by three more commercials for the rest of the hour. KTKT was lucky to air nine records per hour. On Saturdays, it took Ray Lindstrom four and one-half hours to count down the new Top 40 chart. For several years, KTKT carried this load of sales for 18 hours, six days a week, and for only 12 hours on Sunday, due to public-service obligations. (Ray Lindstrom.)

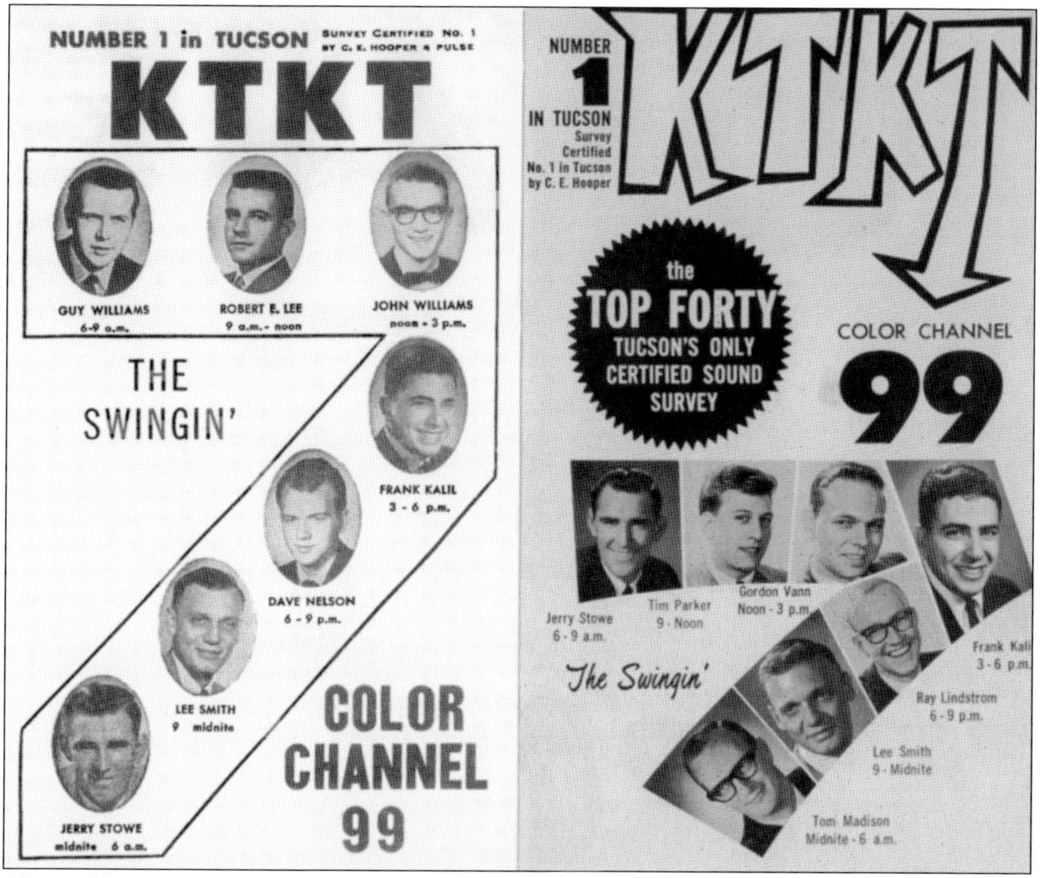

The DJ lineup at left, from April 1962, includes Guy Williams, aka Dave Morehead. He left KIMN in Denver for KTKT, then went on to success in Los Angeles. Robert E. Lee left KTKT for KIMN. John Williams is the KHJ Boss Jock of the future. Dave Nelson moved on to California and Hawaii radio. Lee Smith went to KRIZ in Phoenix and, later, to Seattle. The lineup at right, from May 1964, includes Jerry Stowe, the most memorable morning DJ of the 1960s. Tim Parker died in a tragic automobile accident after leaving Tucson. Gordon Vann returned to Ohio and is now in the Ohio Broadcasting Hall Of Fame. (CJB.)

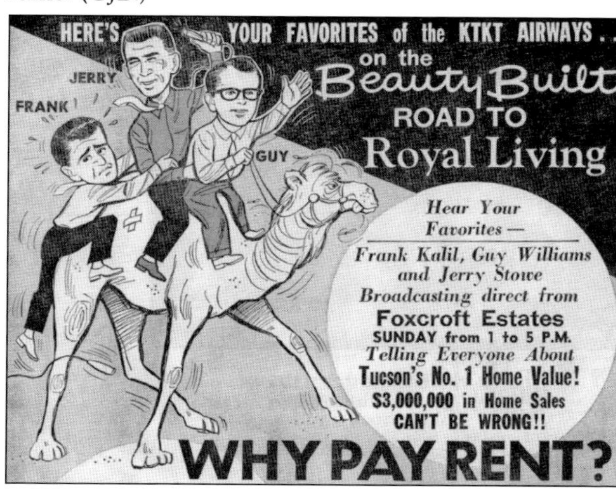

This is a January 1962 advertisement for a remote broadcast for KTKT sponsor Beauty Built Homes. Lily Kay is still going strong in this image, carrying, from left to right, Frank Kalil, Jerry Stowe, and Guy Williams. (Frank Kalil.)

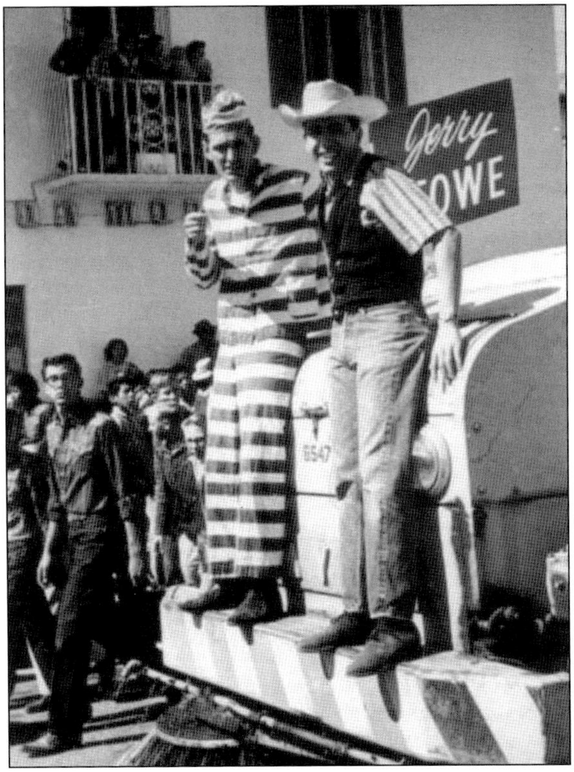

The 1963 Tucson Rodeo Parade included "convict" Jerry Stowe and "sheriff" Frank Kalil (both pictured at left). Ray Lindstrom did evenings for more than three years at KTKT, becoming one of KTKT's most-remembered DJs. He also hosted the popular weekly Top 40 countdown debut every Saturday. All 40 tunes were counted down by Lindstrom in much the same style that Casey Kasem would use years later. Frank Kalil would then debut the new KTKT pick hit right after the 3:00 p.m. newscast following Lindstrom's countdown. Ray Lindstrom became so popular and liked by his audience that he was able to declare himself "King Ray" and don a crown and robe (below). Listeners could write in to Lindstrom at KTKT for free membership cards and a Certificate of Induction into Lindstrom's Legionnaires. (Ray Lindstrom.)

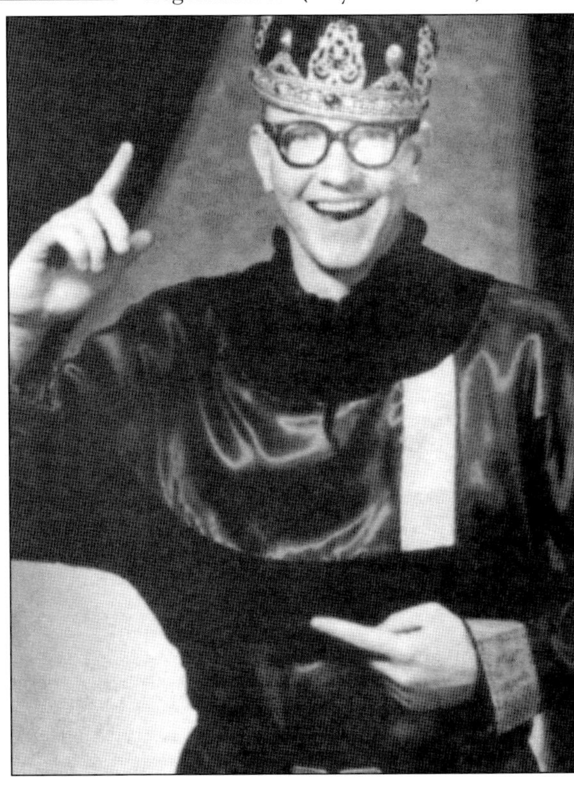

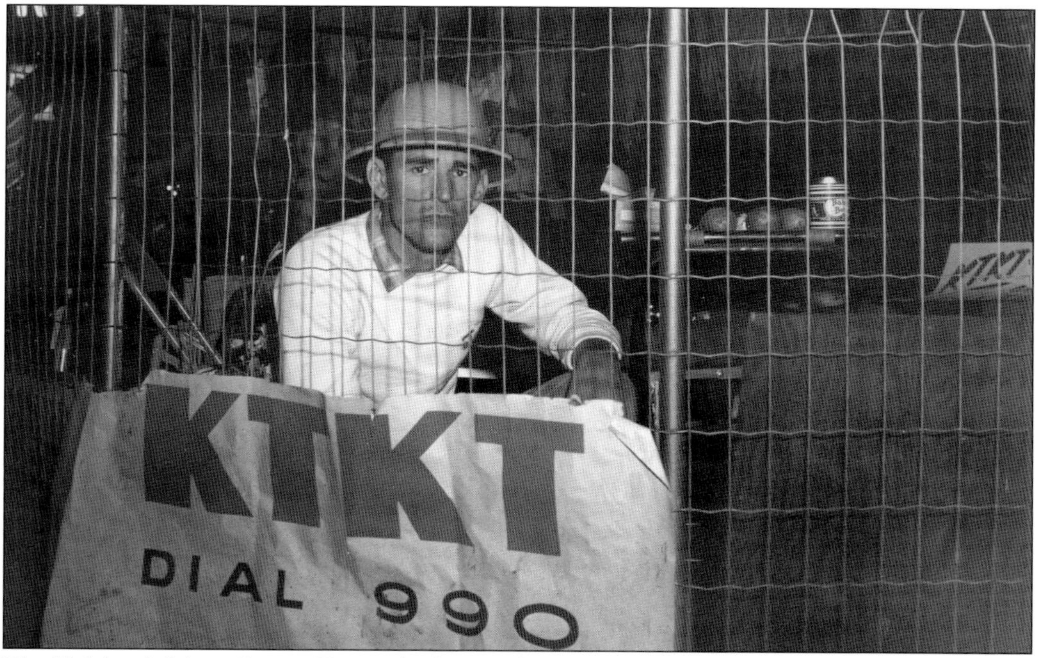

The Tidelands Motor Inn in Tucson had a caged emu in front of the lobby to promote the hotel. In the fall of 1960, the emu came up missing. KTKT DJ Jerry Stowe spent more than a week in the cage (above), promoting both the plight of the missing bird and KTKT. It was later determined that UA students took the emu as a prank that, in the end, failed: the bird died. Stowe, who was a graduate of the Chris Borden DJ School, started at KTKT in April 1960. After doing overnights and middays, Jerry had a long run in morning drive. Similar to Kalil in the afternoon, Stowe often had more than 50 percent Hooper results. Below, Stowe sits on a billboard on busy Speedway Boulevard, promoting the latest KTKT contest. (Jerry Stowe.)

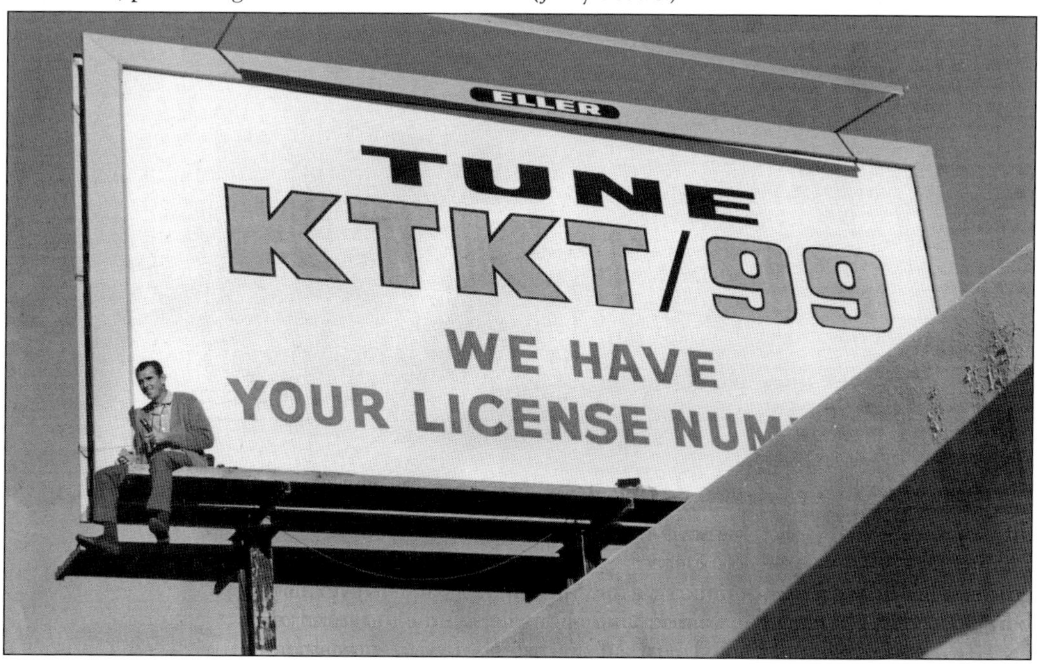

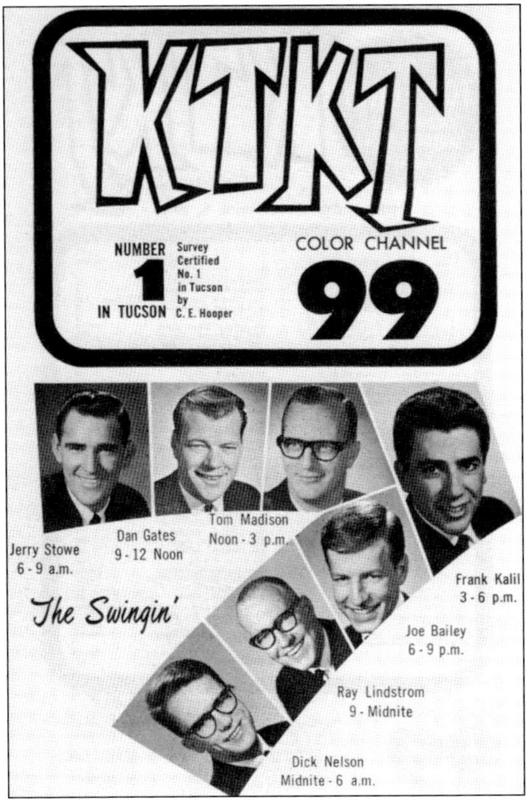

The February 1965 lineup at left shows Dan Gates at 9:00 a.m. Gates has one of the most recognized voices in Tucson, having worked for many local radio and television stations for over 50 years. When joining KTKT in 1965, he took over the PD duties from Jerry Stowe. Tom Madison was a noontime favorite with the ladies while at KTKT. Joe Bailey was an instant hit when he moved from news to be a full-time DJ in 1964. He became program director in 1971. Dick Nelson was the brother of former KTKT DJ Dave Nelson. The coverage map below illustrates that KTKT's 10,000-watt signal reached a good portion of greater Phoenix. The FCC later ordered KTKT to adjust this pattern as a condition to a 1971 sale. (Left, CJB; below, Ray Lindstrom.)

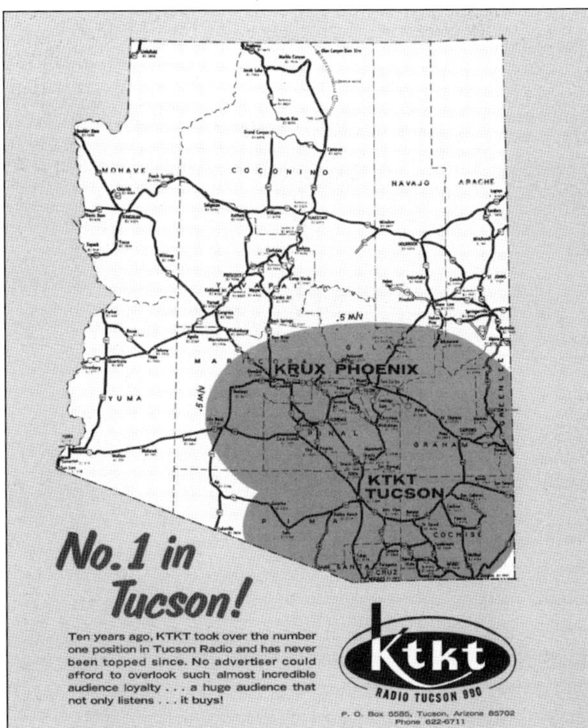

The December 1966 chart at right includes Bob Holiday, who returned for a second tour of duty at KTKT in 1965 from WING in Dayton, Ohio. Tom Wright did well in the evenings, having moved to KTKT from KFIF in early 1966. Lee Poole started doing overnights using his regular voice. But Poole soon developed a very credible Wolfman Jack–type voice and became Johnny Rabbit on weekdays. When doing weekend shifts or reading news as Lee Poole, he used his "regular" voice. Pictured on the left in the below image is funny man Don Hinson, who took over the morning show in November 1967. Hinson went on to stardom in Los Angeles. On the right is the man who took over Kalil's time period, Mike "Boy" Nardone, who previously worked with Bob Holiday at WING in Dayton, Ohio. (CJB.)

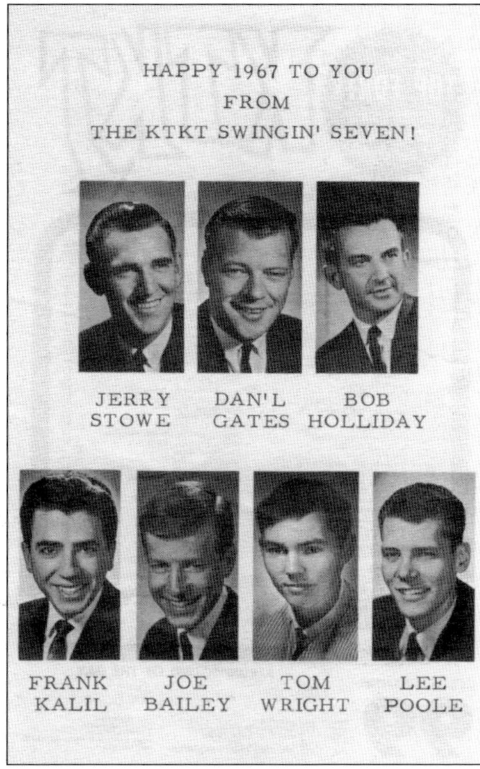

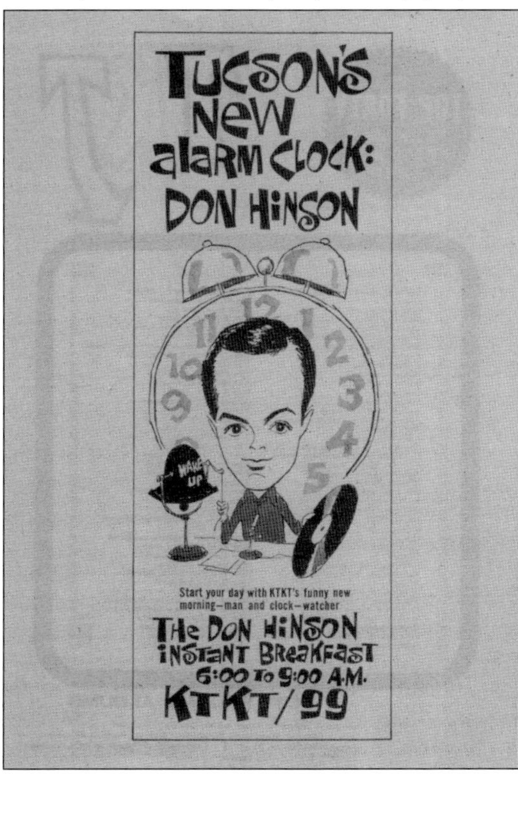

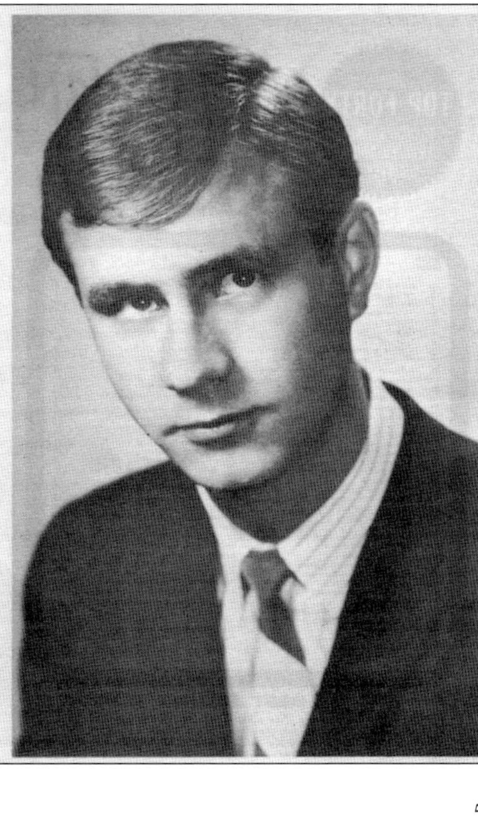

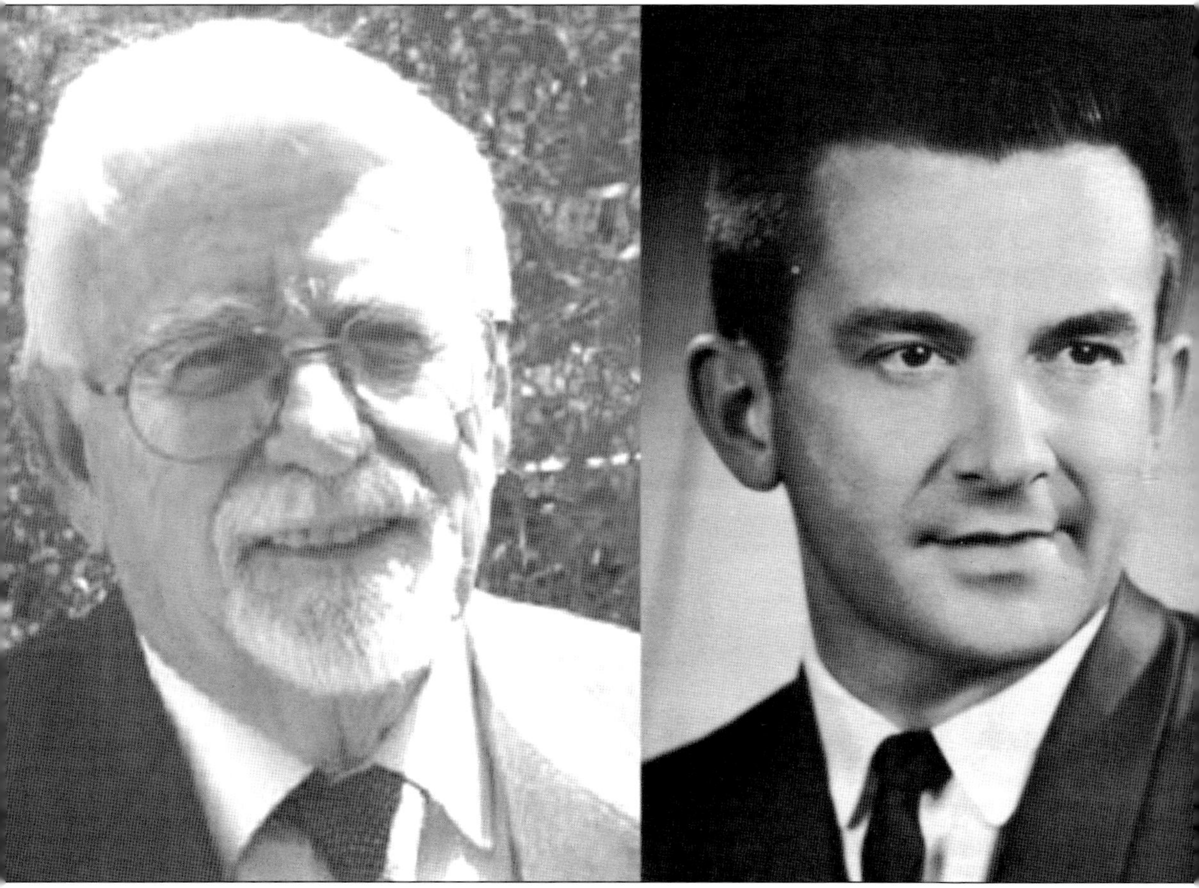

KTKT faced big challenges during the late 1960s. Full-time country KHOS and adult music KCEE, with 24-hour authority granted in 1964, chipped away at KTKT's adult audience. In 1965 and 1966, daytime-only KFIF and KOLD, doing Top 40 evenings, gave the younger listener new choices. The biggest challenge came in June 1967, when KIKX signed on. Due to the heavy KTKT spot load, KIKX stole most of KTKT's younger demo in just six months. KTKT GM Phil Richardson (pictured at left in 2007) once said that KIKX "almost ate KTKT's lunch and would have probably succeeded in doing so if not for a few management blunders." Several changes occurred in 1967. In the spring, KTKT moved to brand-new studios and offices at the West Grant Road transmitter site. Big lineup changes took place near the end of 1967, as Jerry Stowe left the morning show and moved to 9:00 a.m., and PD Dan Gates departed for rival KIKX. Bob Holiday (right) took over PD duties when Gates left. But the most dramatic change for KTKT was Frank Kalil leaving in late 1967 to assume ownership of KAIR. (CJB.)

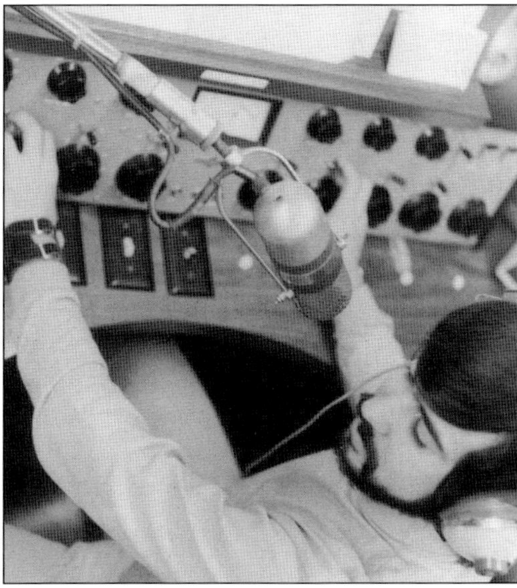

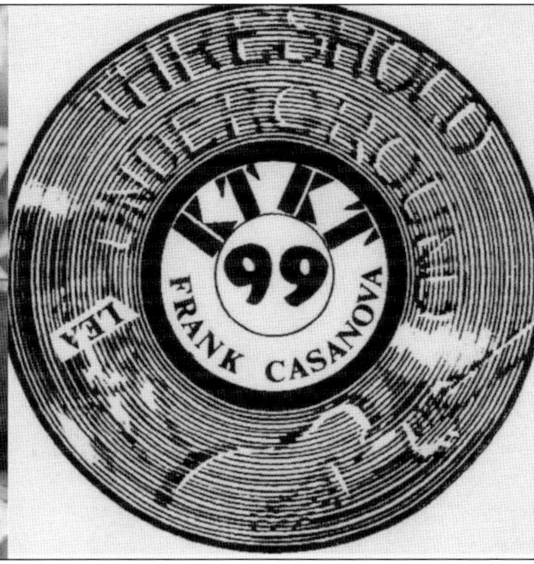

Frank Casanova, at the KTKT controls in 1968 above, earlier worked at KOLD, KFIF, and KIKX. Casanova became very popular with the evening crowd. In 1968, he created and aired one of the earliest "underground" music programs in Tucson for KTKT, *Threshold*. Casanova often hosted concerts, managed and produced local bands, operated his own record label, and owned a record shop that some called Tucson's first "head shop." In 1969, Casanova produced a KTKT documentary film called *Logo* seen in local theaters. Today, he operates a video production studio in California. Pictured at right below are KTKT morning men Lew Jones and Lou Waters. Dan Gates (below left), who returned to KTKT as PD in late 1968 after Bob Holiday departed, promoted the two DJs as a "Lew-Lou" of a morning. Waters later became a CNN newsman. Jones would later head the Jones Radio Network. (Above, Frank Casanova; below, CJB.)

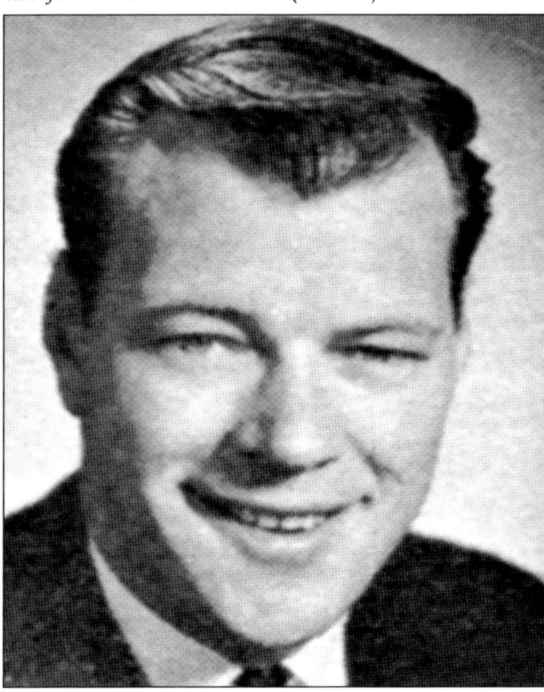

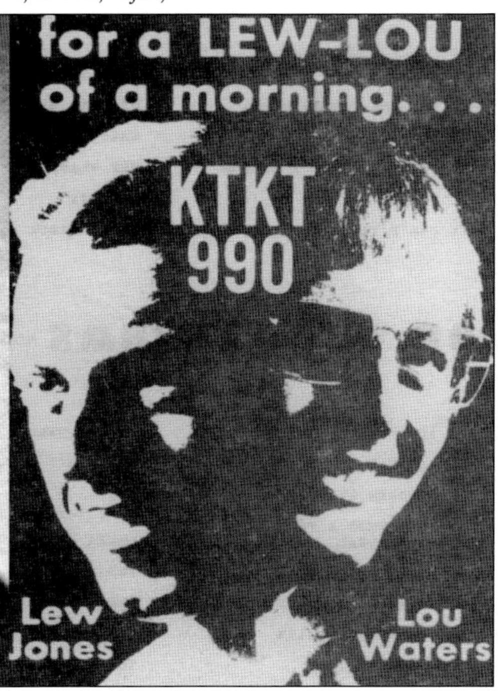

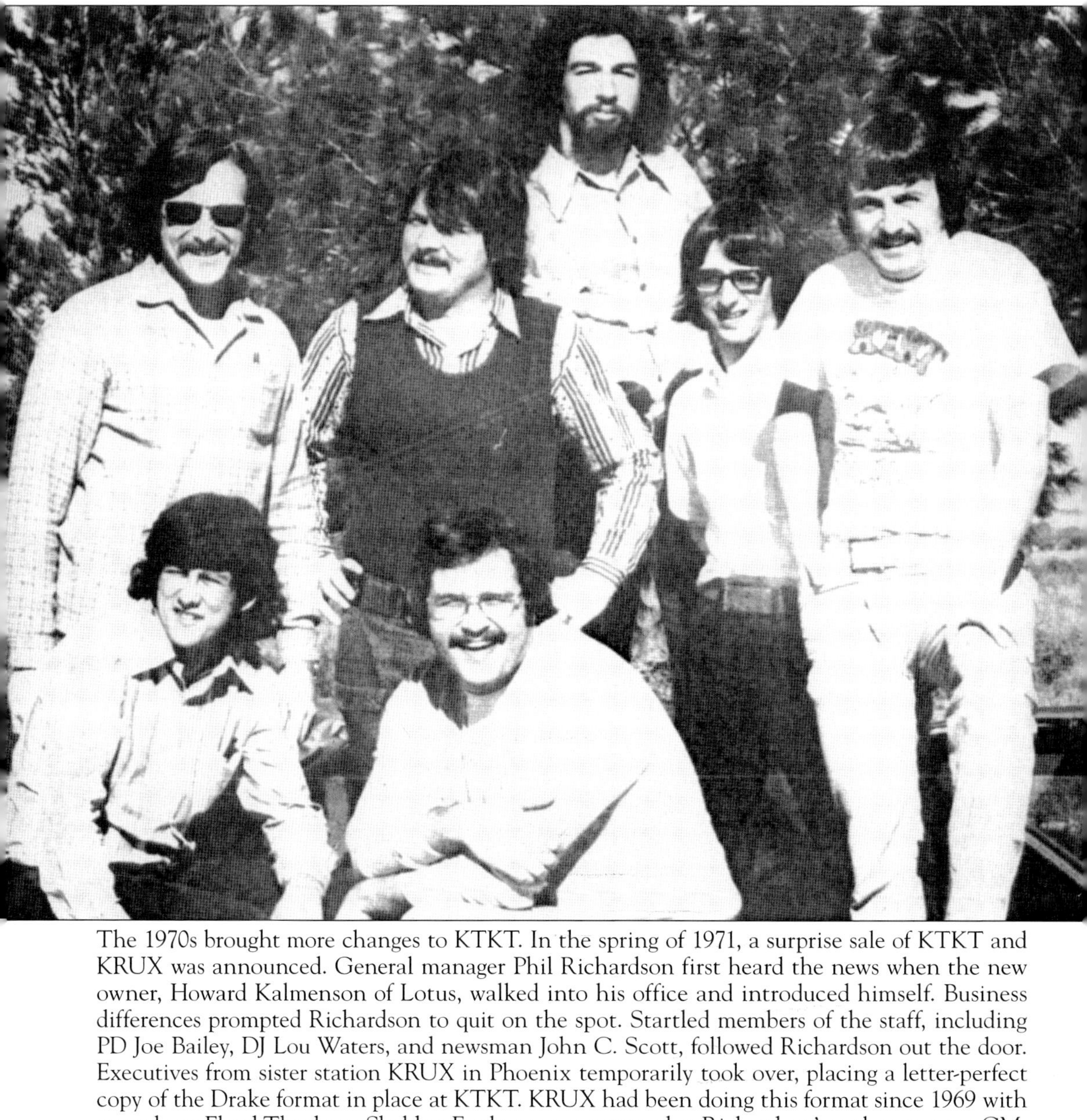

The 1970s brought more changes to KTKT. In the spring of 1971, a surprise sale of KTKT and KRUX was announced. General manager Phil Richardson first heard the news when the new owner, Howard Kalmenson of Lotus, walked into his office and introduced himself. Business differences prompted Richardson to quit on the spot. Startled members of the staff, including PD Joe Bailey, DJ Lou Waters, and newsman John C. Scott, followed Richardson out the door. Executives from sister station KRUX in Phoenix temporarily took over, placing a letter-perfect copy of the Drake format in place at KTKT. KRUX had been doing this format since 1969 with consultant Floyd Thackery. Sheldon Engle was soon named as Richardson's replacement as GM. The sale of KTKT did not close until August 1972. The DJ staff seen here is from the fall of 1972. They are, from left to right, (first row) Rick "Mucho" Morales and Ed O'Brien; (second row) Lee Price, John Mack Flanagan, Ron Wiley, Ed Alexander, and Roger Collins. (Tom Lang.)

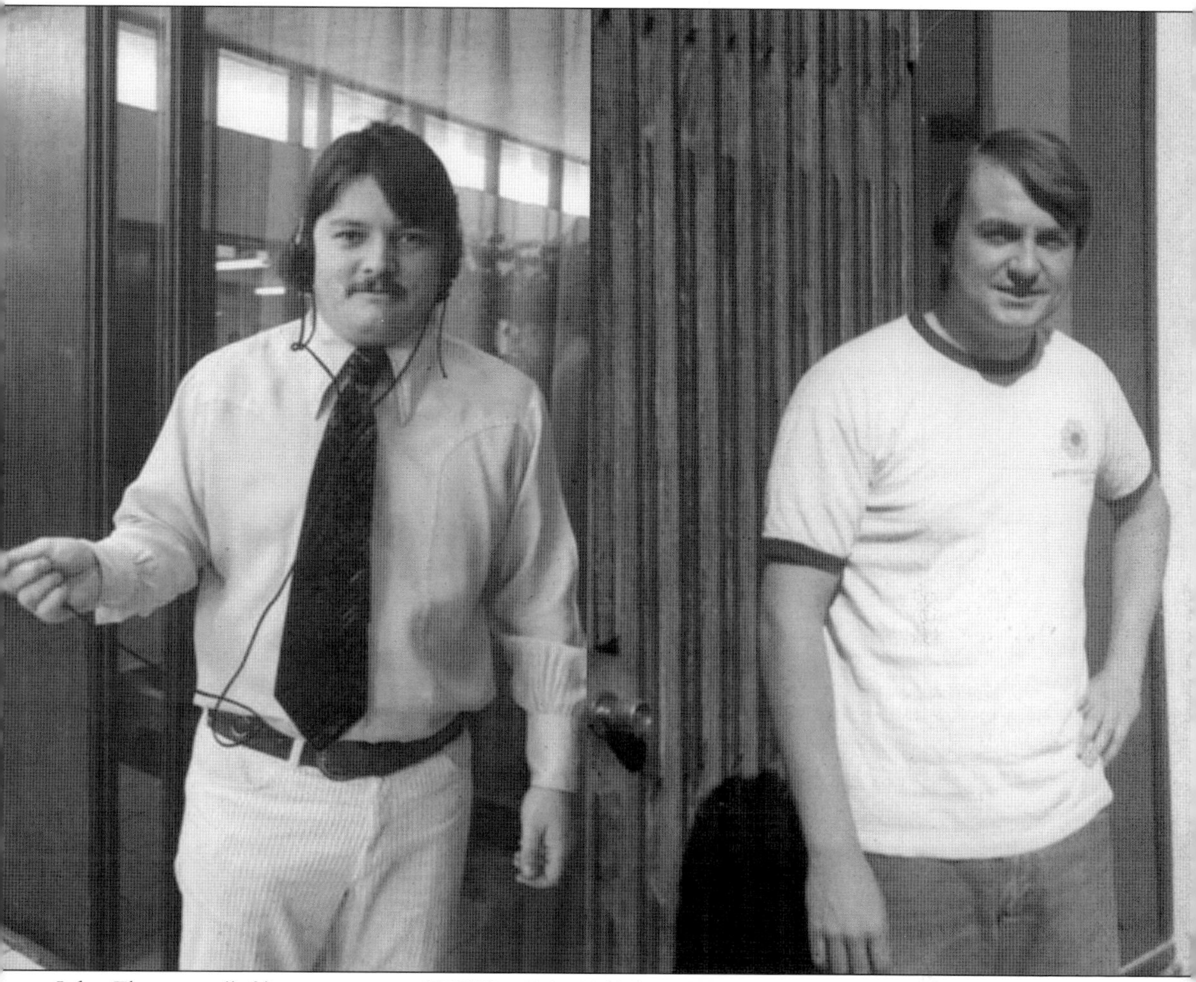

John Flanagan (left) came over to KTKT in July 1969 from KIKX. PD Dan Gates asked Flanagan to use his middle name, Mack, when on the air. Gates soon became very impressed with John's booming, deep voice and smooth delivery, and immediately gave Flanagan the afternoon drive. Flanagan became program director after Joe Bailey left KTKT in 1971. Starting January 1973, John had a short run as PD at KRUX in Phoenix. During September 1973, John went on to enjoy great success at KFRC in San Francisco. Roger Collins (right) came to KTKT during 1972 after a year at KIKX. He worked midday shifts before settling into mornings when Lew Jones left during the fall of 1972. Collins took over as PD when Flanagan went to KRUX. At this time, a rejuvenated KIKX was once again giving KTKT fits. Lotus enlisted famed radio consultant John Rook and hired new GM Mark Swartz to help KTKT develop a fresh sound. Collins worked closely with Rook and stopped the decline. (Tom Lang.)

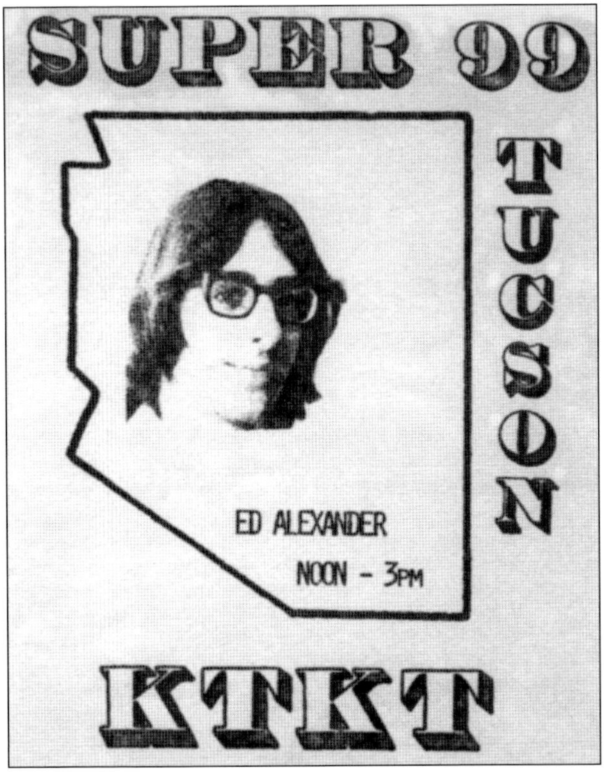

Ed Alexander (left) started at KTKT in 1971 after working previously for KAIR, KWFM, and KHYT. He had a great career run at KTKT, working almost every shift available over a 12-year span. When Roger Collins left KTKT during 1977, Alexander took over the PD duties. In 1979, he oversaw operations at KTKT FM, later KLPX. He left KTKT in 1983 for KCEE. Tom Lang (below) worked several DJ tours at KTKT from 1970 to 1983. In 1984, Chris Haze became operations manager, and KTKT soon dropped live DJs for the Transtar satellite adult contemporary network. Live, local DJs remained on sister station KLPX-FM. On April 10, 1989, KTKT dropped music in order to go all-news. KTKT ran the CNN Headline News radio feed around the clock. After running various mixtures and varieties of news and talk formats, KTKT went Spanish on November 6, 2004. (Left, CJB; below, Tom Lang.)

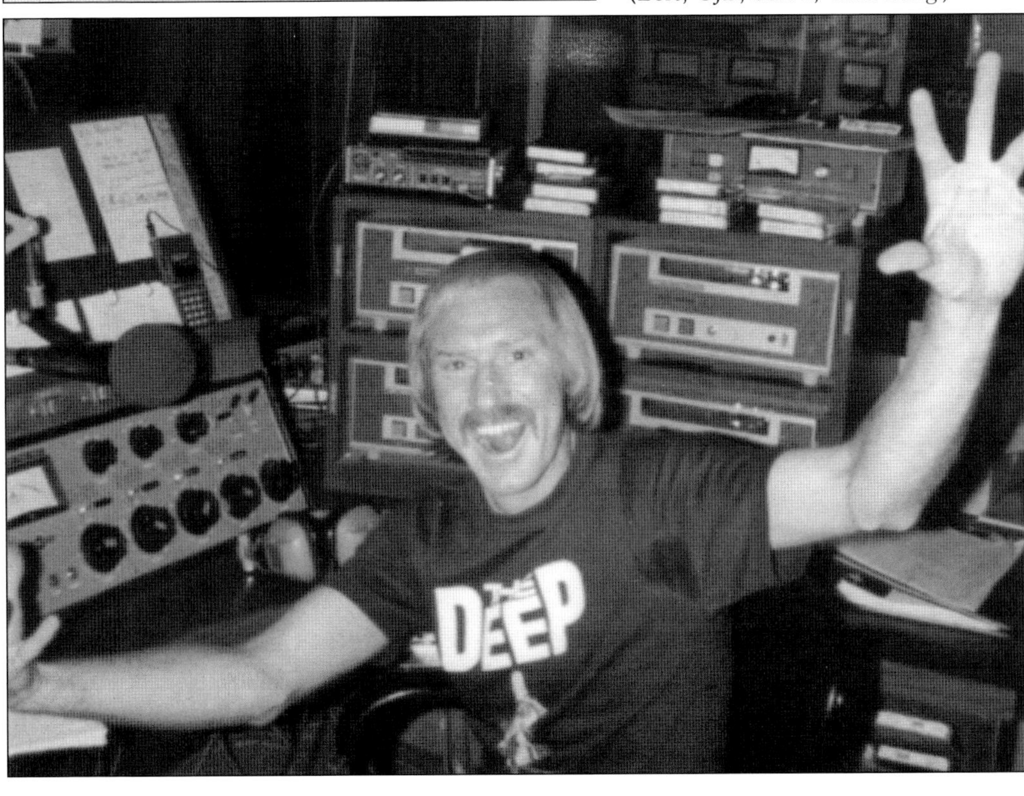

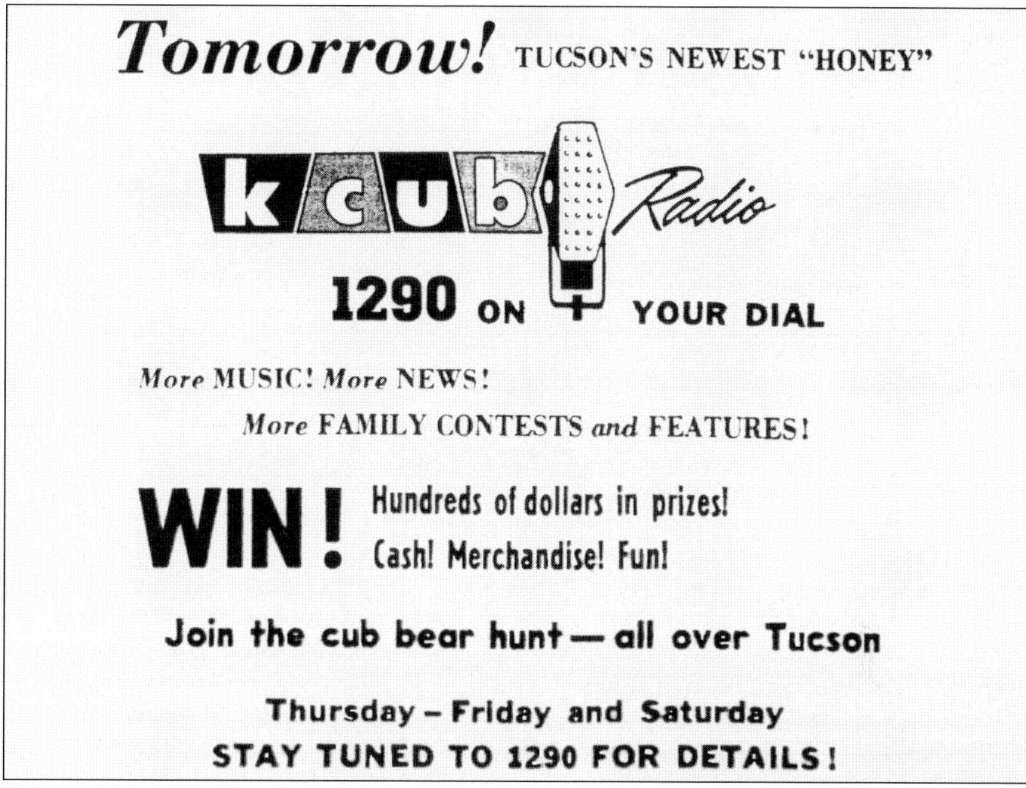

In business, success often breeds imitation. Such was the case after KTKT, then a daytime-only station, took Tucson by storm after introducing Color Radio in May 1957. In 1958, a full-time Tucson station hoped to beat KTKT at its own game. On August 1, 1958, a new owner, Sherwood R. Gordon, took over KVOA-AM 1290. This advertisement announces the new KCUB calls, adopted on September 4, 1958. (CJB.)

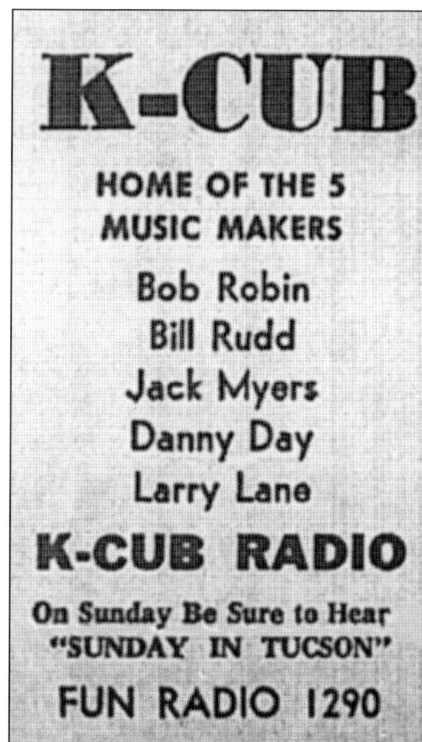

In late September 1958, KCUB debuted a new music format it called "Fun Radio." KTKT veteran Ray Lindstrom remembers KCUB being a "Chicken Rock" operation: "After playing several Top 40 hits in a row, KCUB would break the mood by playing Perry Como." Failing to succeed against a daytime-only competitor, KCUB dropped all Top 40 efforts in March 1960, just before KTKT went to full-time operations. This is an October 1958 advertisement of the KCUB DJ lineup. (CJB.)

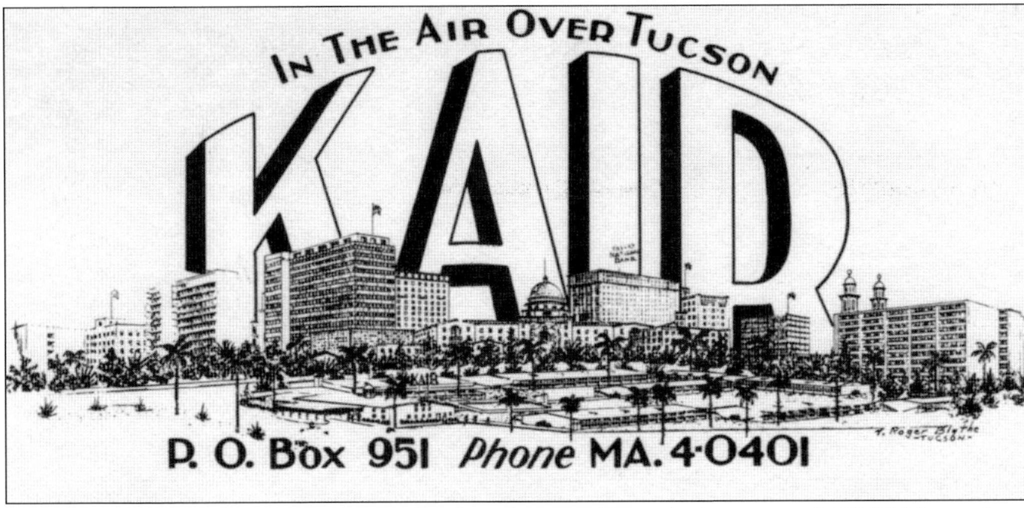

In the spring of 1959, KAIR-AM 1490 also thought it could topple daytime-only KTKT. Launching a Top 40 format under owner Jerry McCauley, KAIR gave up after only two months and switched to country. Russ Jackson, coauthor of this book, worked there for two weeks, filling in for another DJ. He says that the Top 40 format never took off because it sounded too much like "back East" radio. This is the early-1960s KAIR logo. (CJB.)

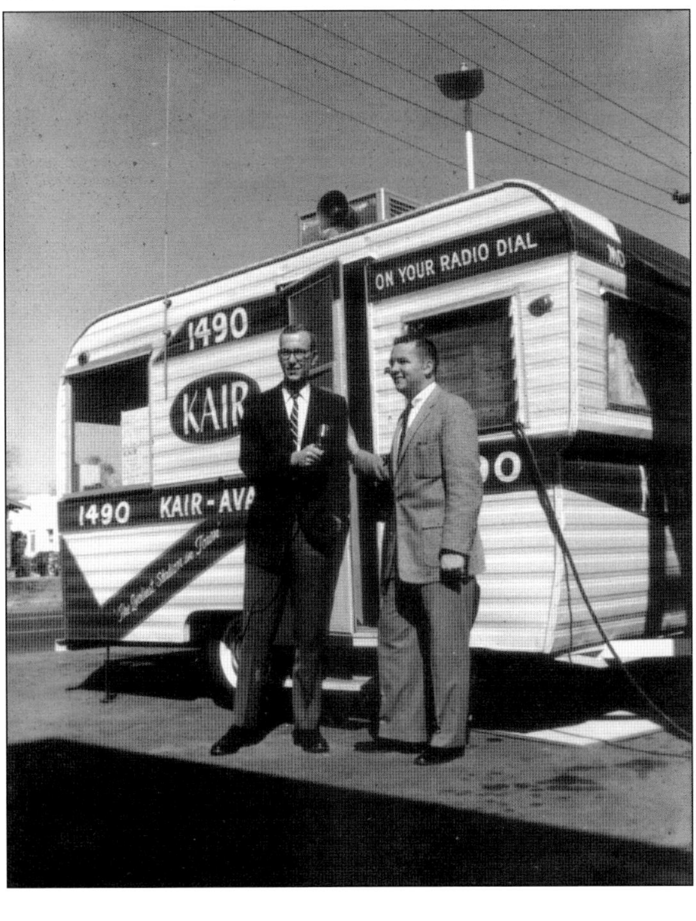

In April 1960, under new owners and different management, KAIR undertook another Top 40 effort. Former KAIR DJ Ray Lindstrom (left) says that he and his partner, Joe Vos (right), "did an overnight brokered program using the KAIR remote unit trailer from a coffeehouse on Fourth Avenue owned by Clyde Battin, himself a rock 'n' roll star as half of the duo Skip and Flip ('It Was I' and 'Cherry Pie')." Battin participated in the program. "Flip" was Gary Paxton. (Ray Lindstrom.)

KAIR

Swingin' 60

RADIO 1490 Tucson

KAIR SWINGING 60 SURVEY #47 FOR WEEK ENDING APRIL 7, 1961

#1	BUT I DO	'Frogman' Henry
2.	BLUE MOON	The Marcels
3.	KATHY	The Innocents
4.	MERRY GO ROUND	Marv Johnson
5.	PLEASE LOVE ME FOREVER	Cathy Jean
6.	ASIA MINOR	Kokomo
7.	TONIGHT MY LOVE TONIGHT	Paul Anka
8.	ONE EYED JACKS	Ferrante & Teicher
9.	UNDERWATER	The Frogmen
10.	DIXIE/GIDGET GOES HAWAIIAN	Duane Eddy
11.	LONELY BLUE NIGHTS	Rosie
12.	SHU-RAH	'Fats' Domino
13.	BE MY BOY	The Paris Sisters
14.	SPANISH HARLEM	Ben E. King
15.	MOTHER IN LAW	Ernie E. Doe
16.	LONESOME ROAD	Mirriam Johnson
17.	SURRENDER	Elvis Presley
18.	I MET HIM ON SUNDAY	The Shirelles
19.	YOU CAN DEPEND ON ME	Brenda Lee
20.	I'VE TOLD EVERY LITTLE STAR	Linda Scott
21.	ON THE REBOUND	Floyd Cramer
22.	HAPPY BIRTHDAY BLUES	Kathy Young
23.	MY THREE SONS	Larry Welk
24.	100 LBS. OF CLAY	Gene McDanials
25.	APACHE	Jorgan Ingman
26.	MATILDA	Cookie & The Cupcakes
27.	RUNAWAY	Del Shannon
28.	BABY BLUE	The Echos
29.	GEE WHIZZ (LOOK AT HIS EYES)	Carla Thomas
30.	PONY TIME	Chubby Checker
31.	PLEDGE OF LOVE	Curtis Lee
32.	BYE BYE BABY	Mary Wells
33.	ALL IN MY MIND	Maxine Brown
34.	BUMBLE BOOGIE	B. Bumble & The Stinger
35.	DON'T BELIEVE HIM DONNA	Lenny Miles
36.	SOME KIND OF WONDERFUL	The Drifters
37.	PONY EXPRESS	Danny & The Jrs.
38.	BONANZA	Al Caiola
39.	TAKE GOOD CARE OF HER	Adam Wade
40.	DON'T WORRY	Marty Robbins
41.	AIN'T IT, BABY	The Miracles
42.	TOP FORTY, NEWS, WEATHER, SPORTS	Marc Dinning
43.	TREES	The Platters
44.	I DON'T WANT TO CRY	Chuck Jackson
45.	AT LAST	Etta James
46.	BOUNTY HUNTER	The Nomads
47.	THE NEXT KISS	Conway Twitty
48.	I WANNA LOVE MY LIFE AWAY	Gene Pitney
49.	ABDUL'S PARTY	Larry Verne
50.	YOU CAN HAVE HER	Roy Hamilton
51.	GLORY OF LOVE	The Roomates
52.	SHY AWAY	Jerry Fuller
53.	THERE'S A MOON OUT TONIGHT	The Capris
54.	WALK RIGHT BACK/EBONY EYES	Everly Brothers
55.	FOOLIN' AROUND	Kay Starr
56.	THE BEST OF EVERYTHING	Frankie Avalon
57.	OH MEIN PAPA	Dick Lee
58.	DEDICATED TO THE ONE I LOVE	The Shirelles
59.	THATS IT, I QUIT	Sam Cooke
60.	THE CLOSER YOU ARE	The Channels

Glen Adams (7 to 12 Midnite)

SAYS:

For Health Sake Wash Face
All Day and Neck All Night

KAIR continued through 1960 and early 1961 with high hopes. A good staff was in place that, at times, included Gary Palant, Marc Avery, Dave Shafer, Johnny Hyde, Dudley Kilburn, and Marshall Darris. Early in 1961, KAIR moved to luxurious studios at the new downtown Ramada Inn. The only problem was money. KAIR owner Josephine Griffith Maye was tiring of the high spending needed to maintain a credible Top 40 format. The station had not turned a profit for her. In mid-May 1961, Tom Wallace Jr., whose father was the original owner of KTKT, was hired as general manager. Wallace promised he could do a gospel music religious format for a lot less money. Within a couple of days being on the job, Wallace told the staff of the format change, to take place the very next day. Needless to say, both employees and listeners were stunned by the sudden change. (CJB.)

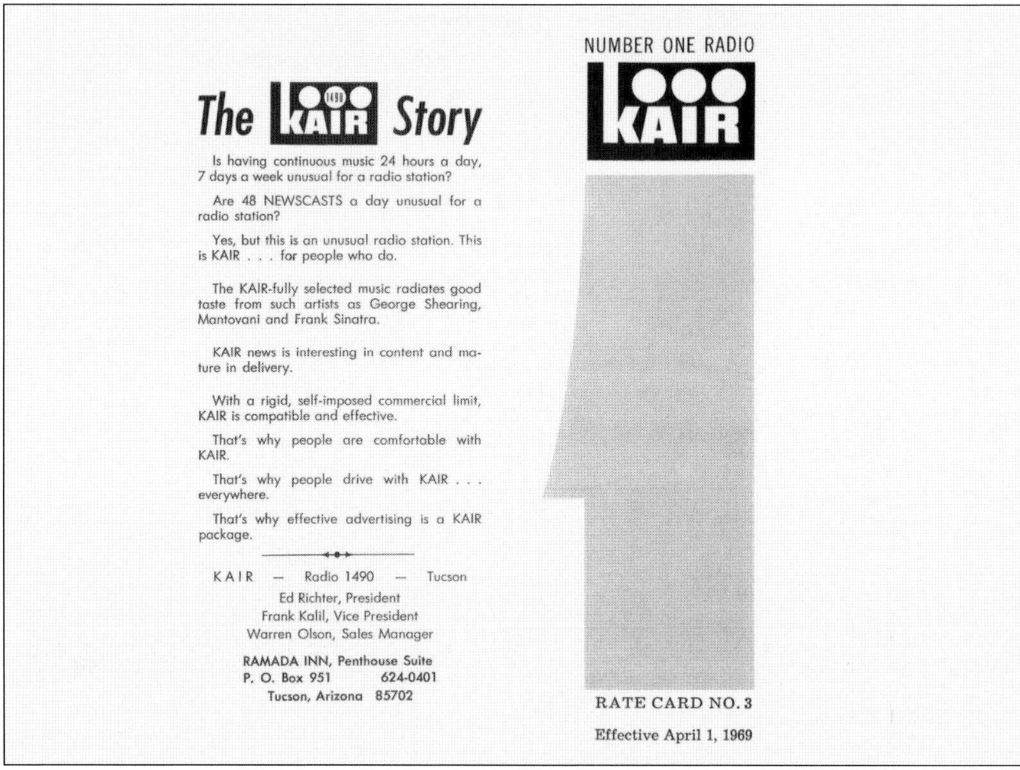

NUMBER ONE RADIO

KAIR

The KAIR Story

Is having continuous music 24 hours a day, 7 days a week unusual for a radio station?

Are 48 NEWSCASTS a day unusual for a radio station?

Yes, but this is an unusual radio station. This is KAIR . . . for people who do.

The KAIR-fully selected music radiates good taste from such artists as George Shearing, Mantovani and Frank Sinatra.

KAIR news is interesting in content and mature in delivery.

With a rigid, self-imposed commercial limit, KAIR is compatible and effective.

That's why people are comfortable with KAIR.

That's why people drive with KAIR . . . everywhere.

That's why effective advertising is a KAIR package.

KAIR — Radio 1490 — Tucson

Ed Richter, President
Frank Kalil, Vice President
Warren Olson, Sales Manager

RAMADA INN, Penthouse Suite
P. O. Box 951 624-0401
Tucson, Arizona 85702

RATE CARD NO. 3

Effective April 1, 1969

In 1967, Frank Kalil formed a business partnership in order to purchase KAIR. Near the end of 1967, Kalil left KTKT to take control of KAIR. Since early 1961, KAIR had featured religious programming. Upon taking control of KAIR, Kalil went to work to give the station a new format. He used the programming skills he developed to perfection at KTKT. Kalil came up with a new adult music format that included many of the same elements found in Top 40 radio, including the best music, good DJs, and lots of local news. KAIR would not be the traditional adult music and news station. Gone was dead air, cluttered program schedules of all varieties, and pointless rambling between the music. Listeners were now greeted with a happy, upbeat presentation that was not jarring. The KAIR brochure above describes the KAIR sound. The KAIR studios in 1972 are shown below. (Above, Frank Kalil; below, Bill Mortimer.)

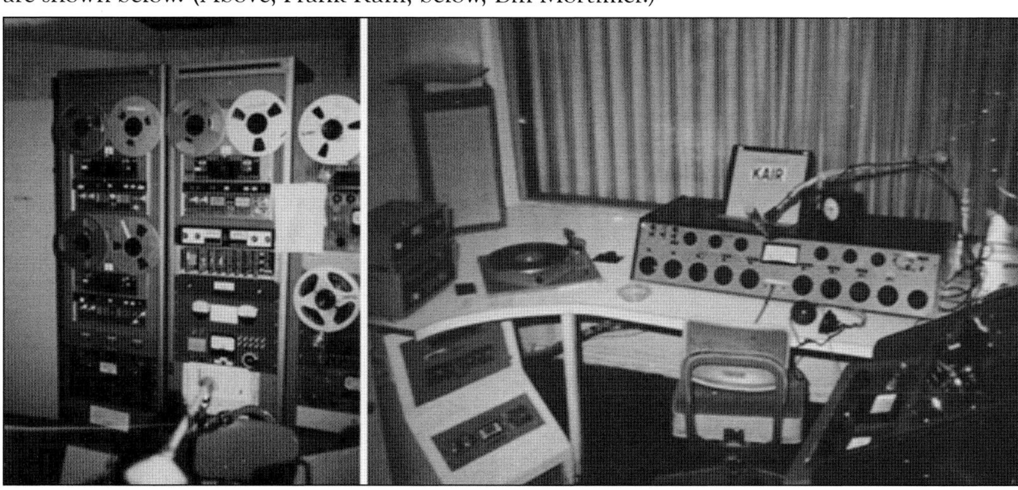

JOHN WASLEY SHOW
"Swinging Soiree"
7:35 - Midnight M - F

☆ ★ ☆ ★ ☆

RICH HEATLEY SHOW
"Heatley on KOLD"
12:35 - 4:30 P.M. M - F

☆ ★ ☆ ★ ☆

JERE LYNN SHOW
"Teen-Scene Reporters"
2:05 - 3:00 P.M. Sat.

☆ ★ ☆ ★ ☆

JIM ARNOLD SHOW
"Weekend Warrior"
Sat. and Sun. Nights

☆ ★ ☆ ★ ☆

KOLD DIAL 1450

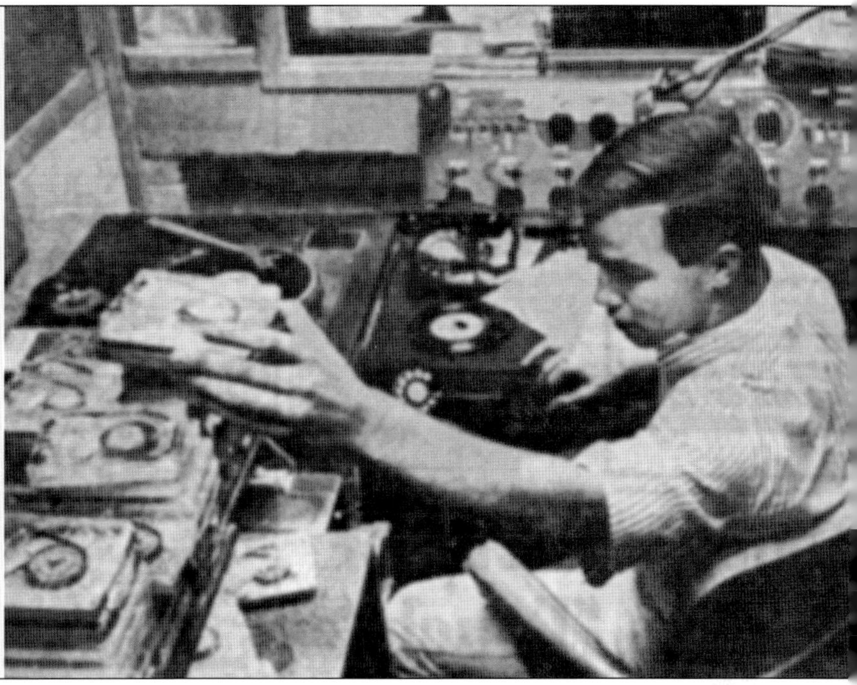

In the spring of 1965, Tucson listeners found a new local Top 40 evening alternative from a very unlikely source, AM 1450 KOLD. Featuring daytime CBS radio stars Arthur Godfrey and Art Linkletter, several long news blocks daily, and older adult standards during music programs, KOLD was probably regarded as a very stodgy radio station by anyone under age 30. Imagine the surprise when regular listeners tuned in evenings expecting to hear "Sunset Serenade" and instead were greeted by music from groups such as the Beatles, the Byrds, and the Turtles. The man behind this music was UA law student John Wasley, who became known simply as "The Waz." Wasley was very much on top of the current music happenings and was often ahead of the curve playing new hits. The big attraction to his program was his laid-back style. Using a delivery laced with sarcastic humor and a dose of rebellion, Wasley was viewed as the ultimate "Mr. Cool." His on-air presentation and style was a sneak preview of the free-form progressive rock FM format of the 1970s. (CJB.)

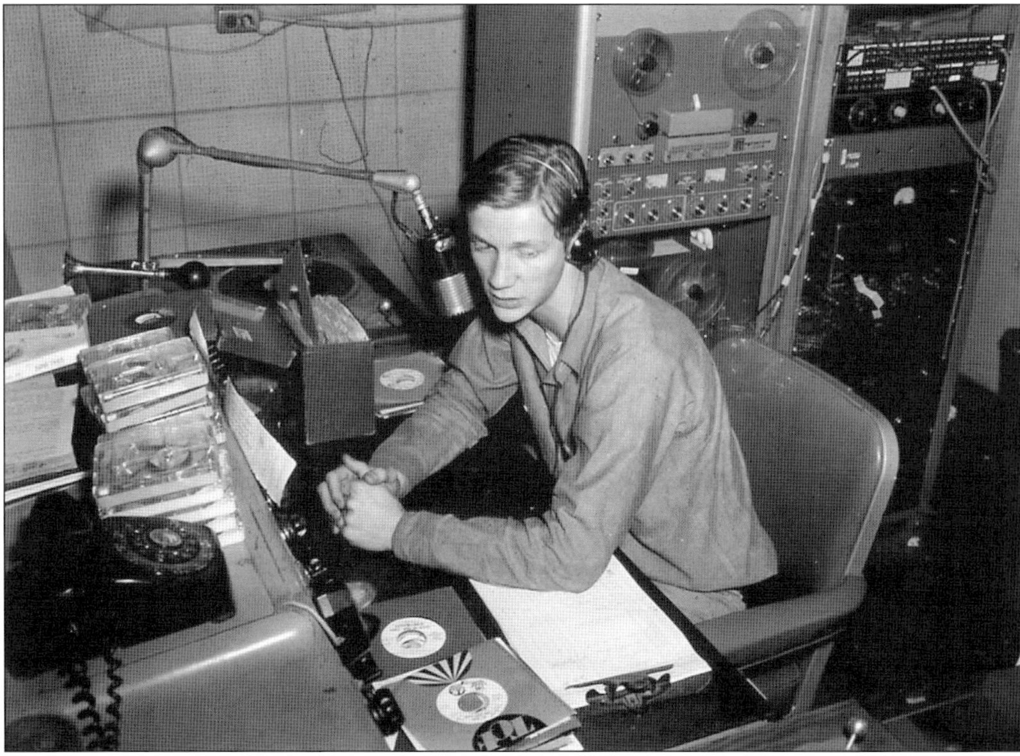

Jim Arnold started at KOLD working the evening shifts on Saturday and Sunday, playing the Top 40 hits heard all week during John Wasley's program. After KOLD, Arnold moved on to do more Top 40 at KTKT and, later, at KIKX. During the 1970s, "Sunny Jim" enjoyed success doing mornings at Tucson country leader KCUB. In later years, he moved on to even more success in television station management at KOLD-TV and KGUN-TV. (Jim Arnold.)

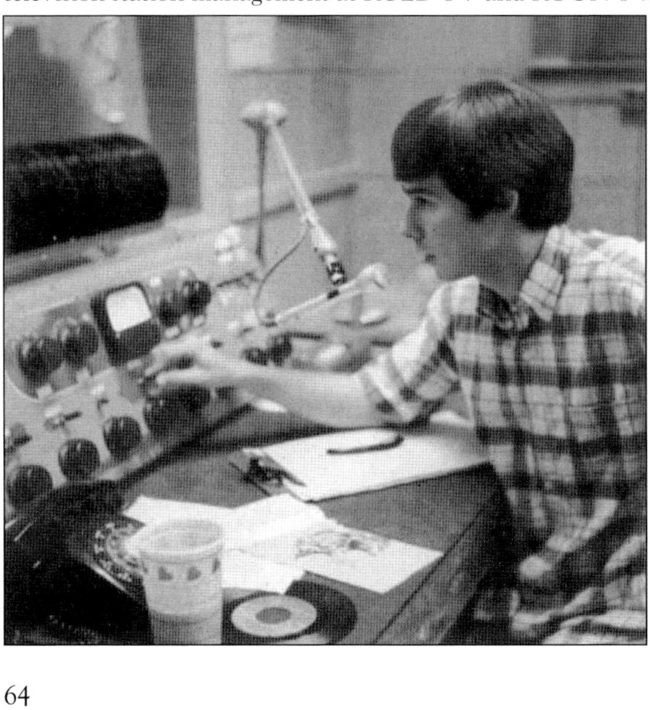

After Jim Arnold left KOLD, Frank Casanova came on board to do late evenings and weekend shifts. Casanova soon became a nighttime favorite. In early 1967, Casanova left KOLD for weekends at KFIF, then stayed on after that station morphed to KIKX. In late 1967, KTKT came calling, and Casanova moved over to do evenings at that station. (Frank Casanova.)

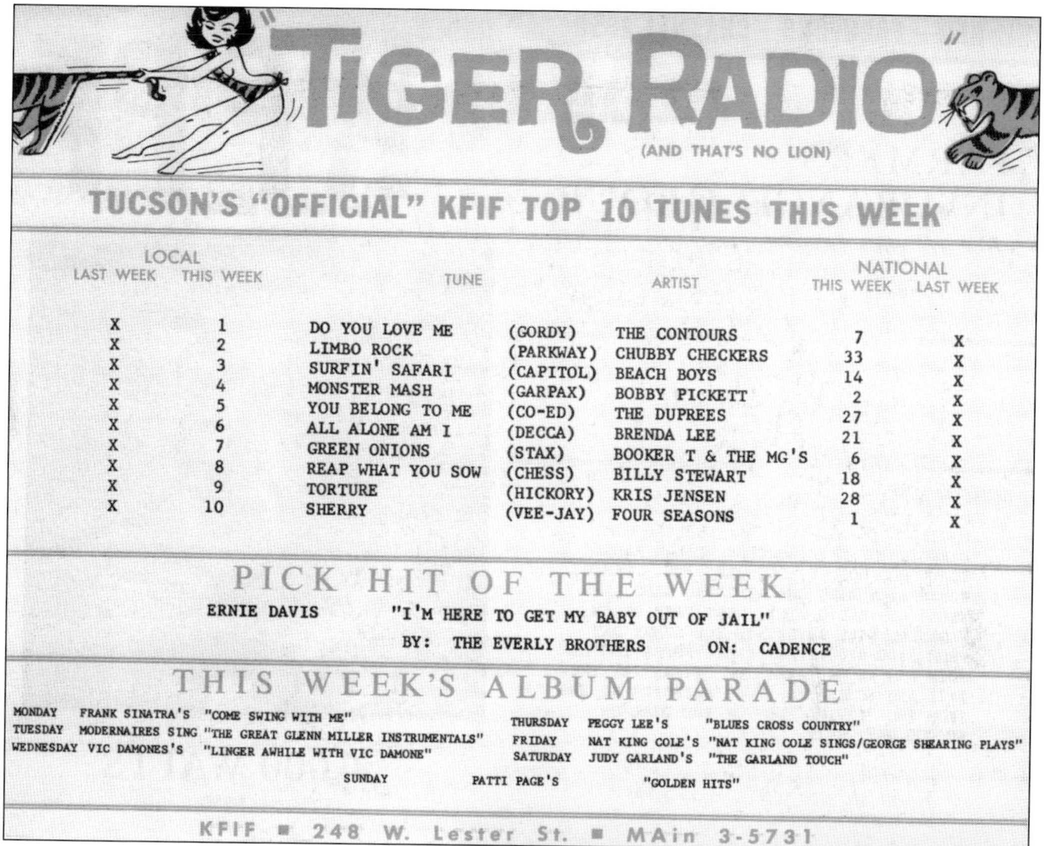
"TIGER RADIO" (AND THAT'S NO LION)

TUCSON'S "OFFICIAL" KFIF TOP 10 TUNES THIS WEEK

LOCAL LAST WEEK	LOCAL THIS WEEK	TUNE		ARTIST	NATIONAL THIS WEEK	NATIONAL LAST WEEK
X	1	DO YOU LOVE ME	(GORDY)	THE CONTOURS	7	X
X	2	LIMBO ROCK	(PARKWAY)	CHUBBY CHECKERS	33	X
X	3	SURFIN' SAFARI	(CAPITOL)	BEACH BOYS	14	X
X	4	MONSTER MASH	(GARPAX)	BOBBY PICKETT	2	X
X	5	YOU BELONG TO ME	(CO-ED)	THE DUPREES	27	X
X	6	ALL ALONE AM I	(DECCA)	BRENDA LEE	21	X
X	7	GREEN ONIONS	(STAX)	BOOKER T & THE MG'S	6	X
X	8	REAP WHAT YOU SOW	(CHESS)	BILLY STEWART	18	X
X	9	TORTURE	(HICKORY)	KRIS JENSEN	28	X
X	10	SHERRY	(VEE-JAY)	FOUR SEASONS	1	X

PICK HIT OF THE WEEK

ERNIE DAVIS "I'M HERE TO GET MY BABY OUT OF JAIL"

BY: THE EVERLY BROTHERS ON: CADENCE

THIS WEEK'S ALBUM PARADE

MONDAY FRANK SINATRA'S "COME SWING WITH ME"
TUESDAY MODERNAIRES SING "THE GREAT GLENN MILLER INSTRUMENTALS"
WEDNESDAY VIC DAMONES'S "LINGER AWHILE WITH VIC DAMONE"
THURSDAY PEGGY LEE'S "BLUES CROSS COUNTRY"
FRIDAY NAT KING COLE'S "NAT KING COLE SINGS/GEORGE SHEARING PLAYS"
SATURDAY JUDY GARLAND'S "THE GARLAND TOUCH"
SUNDAY PATTI PAGE'S "GOLDEN HITS"

KFIF ■ 248 W. Lester St. ■ MAin 3-5731

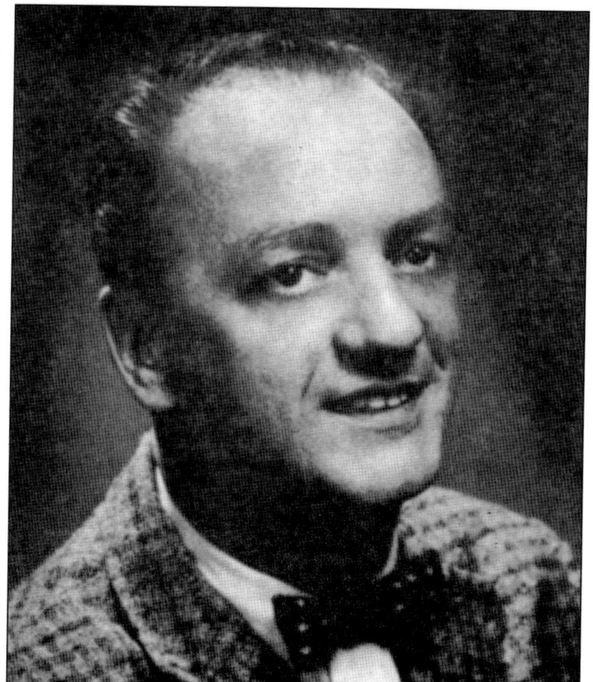

When KFIF 1550 signed on the air, on August 15, 1961, it became Tucson's ninth AM station and Arizona's first 50,000-watt station. KFIF could broadcast only during daylight hours, as KTKT had done in its first years on 990. KFIF debuted using an adult format. On October 15, 1962, KFIF went to a Top 40 format, using a four-man air staff featuring Jack Jackson, Marty Marshall, Doug Douglas, and Ernie Davis. KFIF offered Tucson a Top 40 choice. KFIF published a weekly top 50 chart, news was heard at 55 minutes past the hour, and requests and dedications were aired on the weekends. After a nearly two-month effort, KFIF went silent on December 17, 1962. It returned to the air with an adult format in early 1963. Shown above is the top-10 portion of the debut KFIF Tiger Radio chart. Pictured at right is Ernie Davis, KFIF Tiger Radio program director. (CJB.)

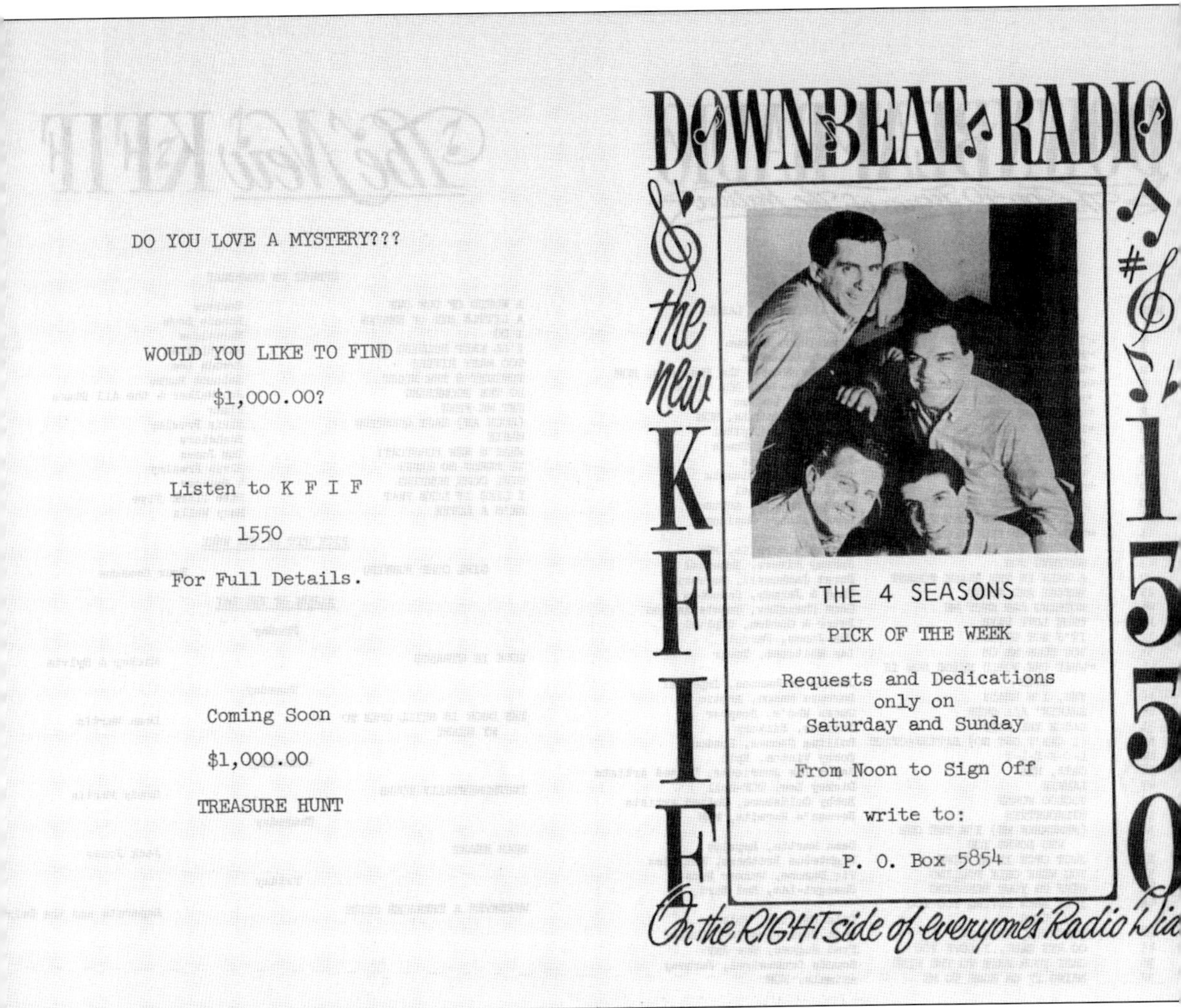

In 1964, KFIF dropped the adult "Variety Radio" format and returned to Top 40. The slogan was "Downbeat Radio," and KFIF started printing a weekly chart. Without much other publicity, KFIF offered a lot of music but little else. A two-man air staff consisting of Bill Steadman and Johnny Holiday split the KFIF broadcast day. In order for each to have a day off, Jay Bundy, the chief engineer, would do a weekend shift. On the weekdays, KFIF aired hourly reports of entertainment news from veteran celebrity reporter Jimmie Fidler. Pictured here is the cover of the June 26, 1964, KFIF Downbeat chart. (CJB.)

On September 1, 1965, KFIF was taken over by new owner John Walton of Texas. Walton put a lot of money into KFIF, totally revamping the station. Changes were made both on and off the air. KTKT now had a vigorous opponent for its Top 40 crown. The biggest problem for KFIF at this time was being limited by the FCC to broadcasting only in the daytime. This same problem had plagued KFIF's 1962 "Tiger Radio" effort. This is the cover of a 1965 KFIF promotional brochure. (CJB.)

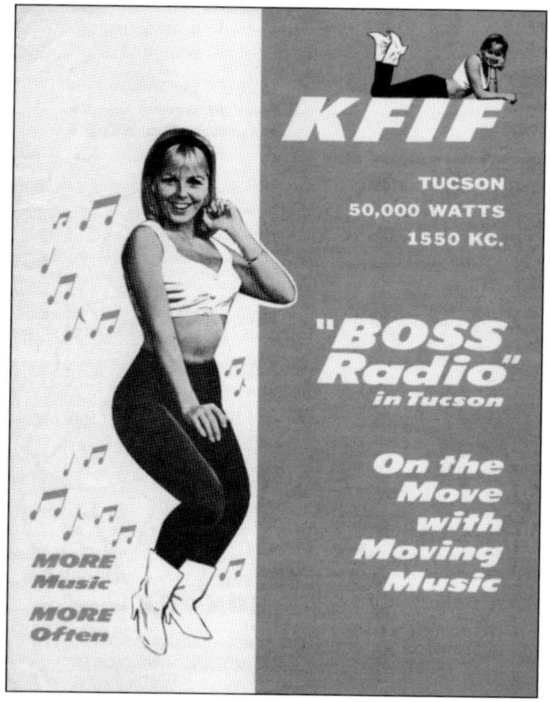

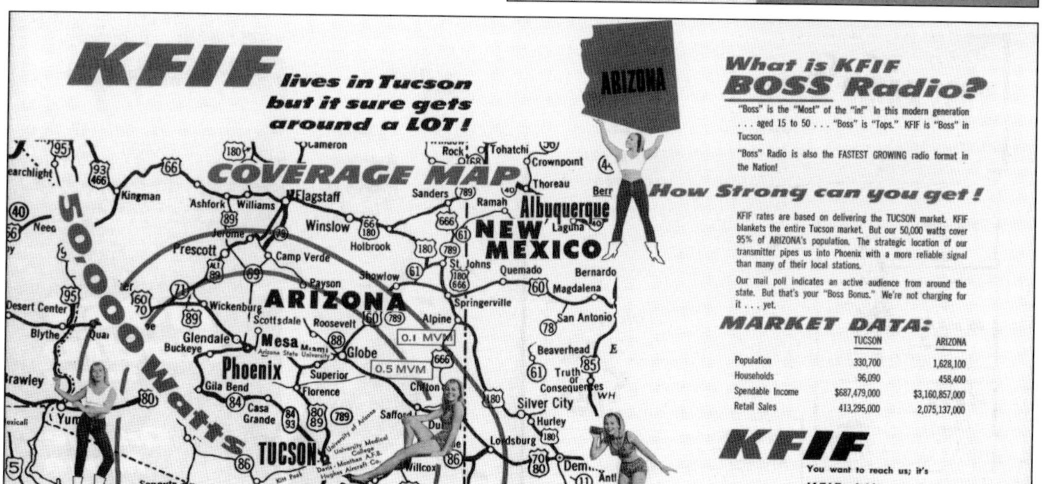

Changes on the air at KFIF included using the services of California radio consultant Ted Randle to debut an exciting new entrant to Tucson, the "Boss Radio" format. This format was a close copy of America's hottest format of the era, heard at KHJ in Los Angeles and at other stations consulted by Bill Drake, the original format developer. Ken White, Earl Trout, Jimmy Love, and Tommy Wright soon took over the weekday air staff. Jay Bundy, still the chief engineer, became a morning newsman. A new, shorter chart, called the "Boss 30," was published weekly. This is a coverage map from the 1965 KFIF promotional brochure. (CJB.)

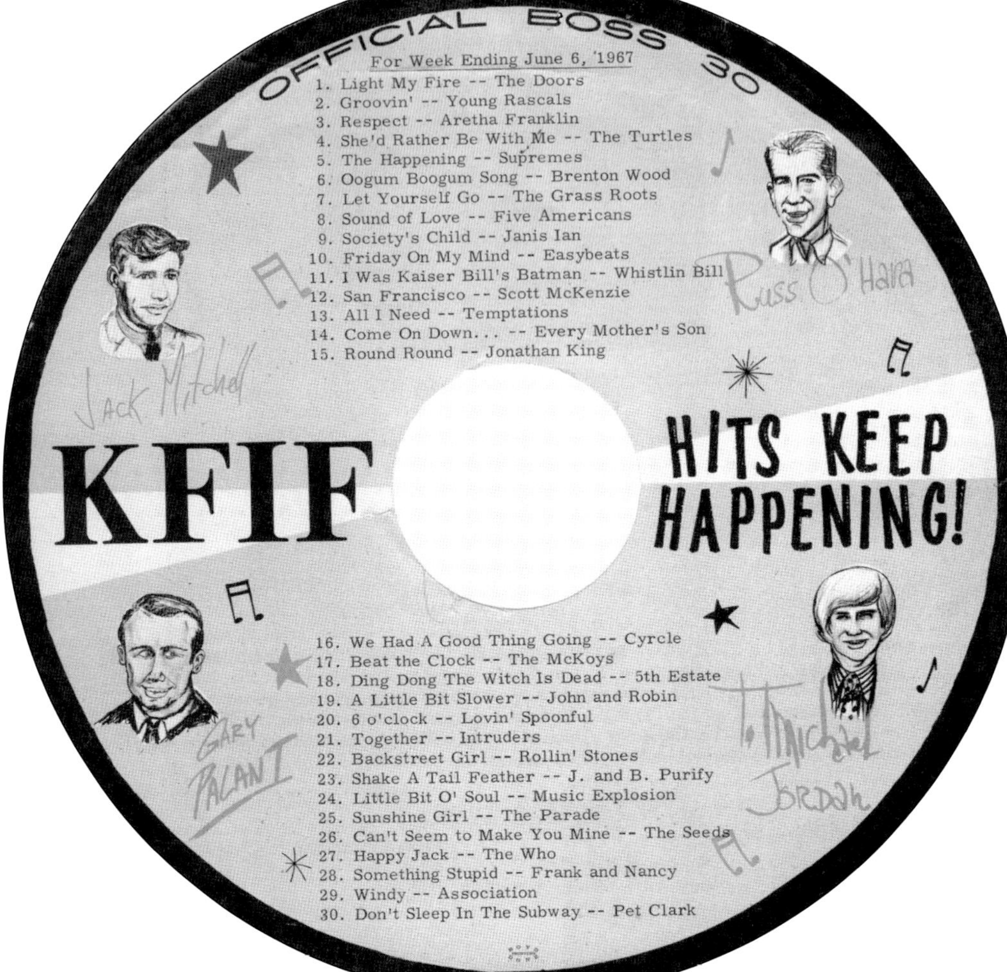

During the spring of 1966, KFIF air-staff changes included Mark Parr taking over mornings and T. Michael Jordan doing afternoons. By fall, Tom Lane and Russ O'Hara were doing the midday shifts. Jack Mitchell took over news in late 1966. KFIF ran several high-profile contests and promoted many concerts that year. A weekly newspaper-type magazine, *The KFIF Boss Green Sheet*, now incorporated the weekly chart. Despite a good-sounding format that was executed reasonably well, KFIF was never able to gain a lot of success battling KTKT. By late 1966, KFIF realized that a new game plan was needed. Owner John Walton devised a plan to donate KFIF to the University of Arizona in order to gain FCC approval to purchase another Tucson station. Walton's new weapon to dethrone KTKT would be what was then AM 580 KTAN, a station with full-time broadcast authority. Shown here is the final KFIF lineup from the last published KFIF chart. (CJB.)

KFIF AM 1550 signed off forever when KIKX signed on for the very first time at 3:00 p.m. on Saturday, June 10, 1967. The first record heard on KIKX was the Fifth Dimension's smash "Up, Up and Away." KIKX continued to play back-to-back music with no DJs and little interruption all weekend. The live regular DJ format began at 6:00 a.m., Monday, June 12, when the Lovin' Spoonful's "Six 'O'Clock" signaled the start of Tucson's great Top 40 war. For the first time since May 1961, KTKT now had full-time competition. This is a KIKX promotional advertisement from June 1967. (Roger Galloway.)

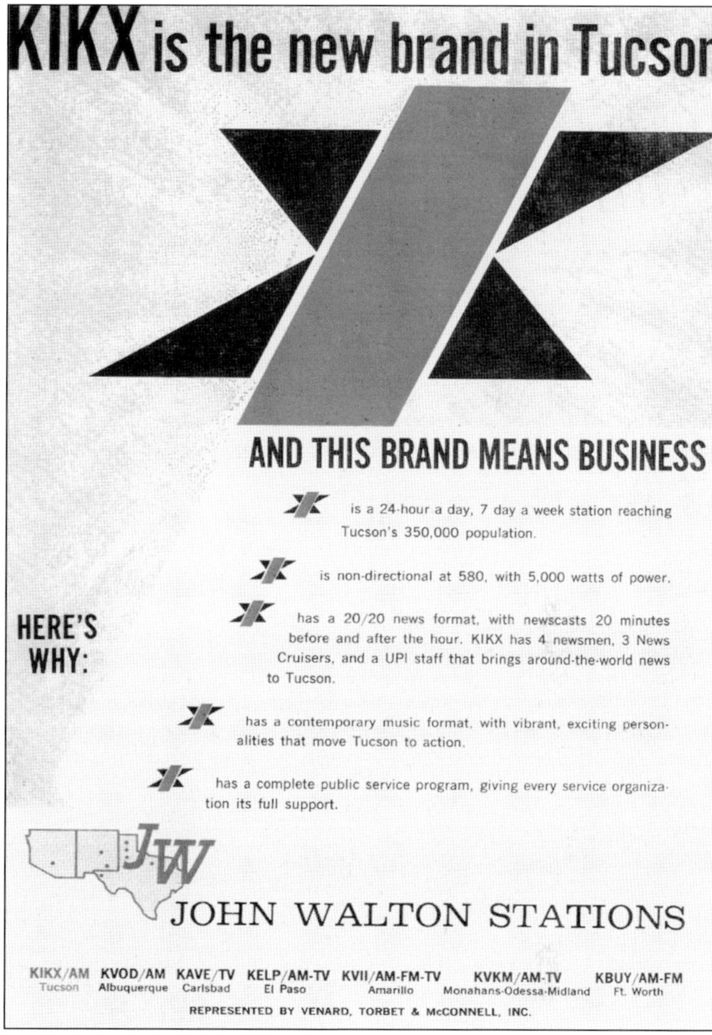

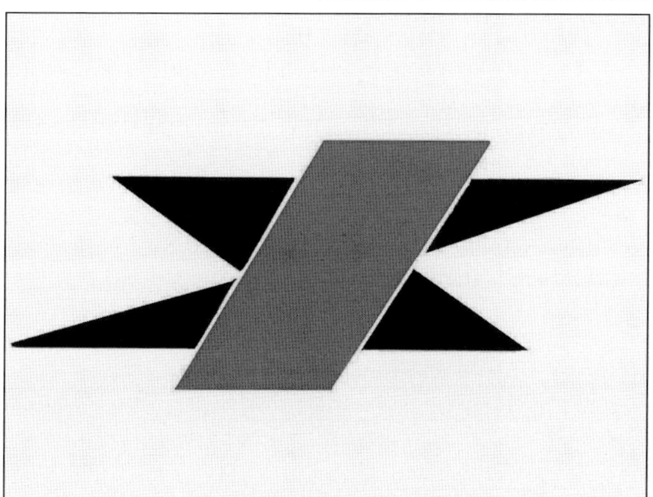

Over-the-air talent was an area in which KIKX hoped to excel. Several fresh voices were added to some holdover KFIF staff to create an exciting sound. Shortly after sign-on, Roger Galloway, who previously worked at KTKT, became the director of the KIKX news department. Other news people at start-up included Ron Robbins, Lon Ritzo, Bob Savage, and Casey Jones. The news department started out with two mobile units. Shown here is the original KIKX logo. (CJB.)

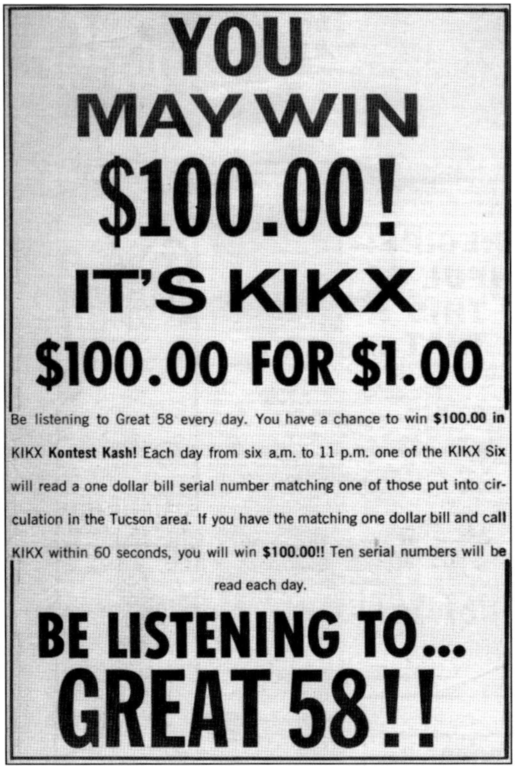

During its first year on the air, KIKX featured many contests. KIKX probably ran more contests over a year than most stations ran in a decade. Often, more than one contest each day could be heard. Prizes ranged from great to not so great. Some contests required a phone call to the station, while others needed a written entry. One of the most memorable KIKX contests ran during the summer of 1968. The station promised to give a car to someone each week. Excitement ran high when the KIKX Car-a-Week contest was announced, but excitement sputtered when it was discovered the cars given away were used cars from a local dealer. For a short time, KIKX was Tucson's contest king. The advertisement at left announces the KIKX Kash giveaway. Below, KIKX promotes its 1968 car-a-week giveaway. (CJB.)

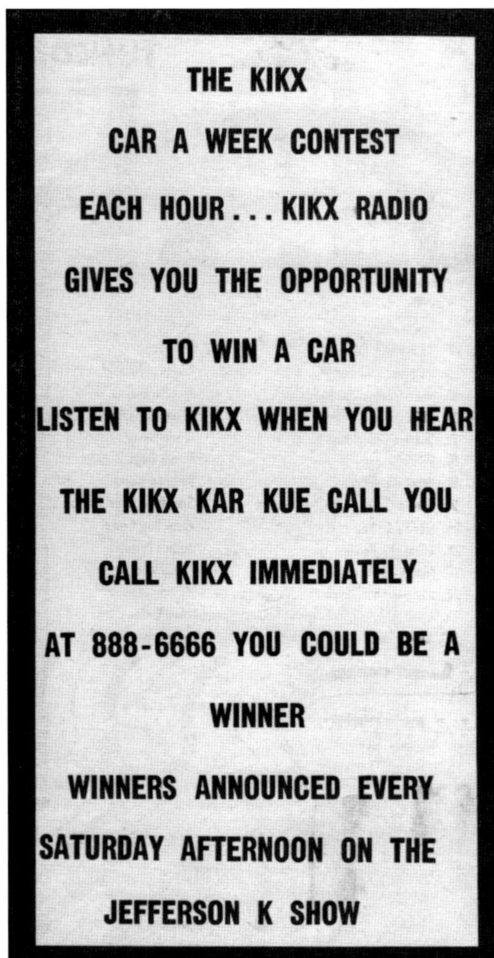

KIKX not only ran many contests, it also sponsored or promoted many concerts during its first year. Some of the acts KIKX brought to Tucson included the Doors, the Young Rascals, Buffalo Springfield, the Box Tops, the Strawberry Alarm Clock, the Buckinghams, and Eric Burdon and the Animals. One of the biggest shows in Tucson during 1968 was a new fellow just starting to heat up, Jimi Hendrix. Many other notable acts of the era also appeared in Tucson at this time. The most utilized venues were Hi Corbett Field and the VIP Club on Speedway. Local bands often opened for the big names. Pictured at right is an advertisement from the *KIKX Go* newspaper for the Young Rascals August 1968 concert. The below image, also from *KIKX Go*, touts the October Doors concert. (CJB.)

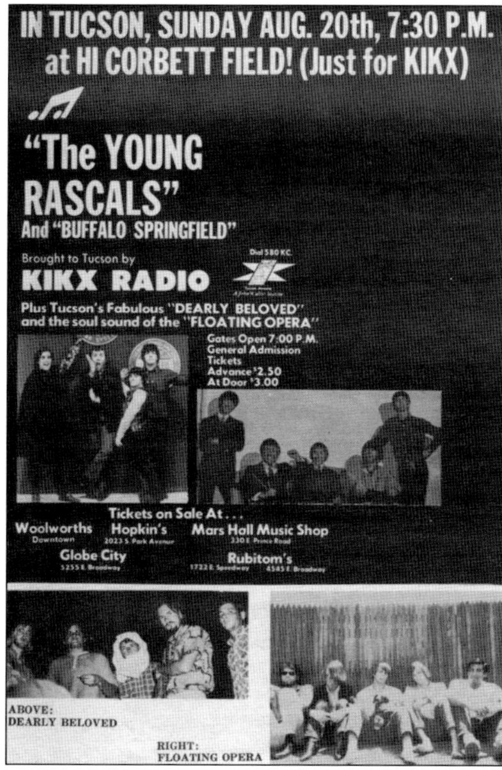

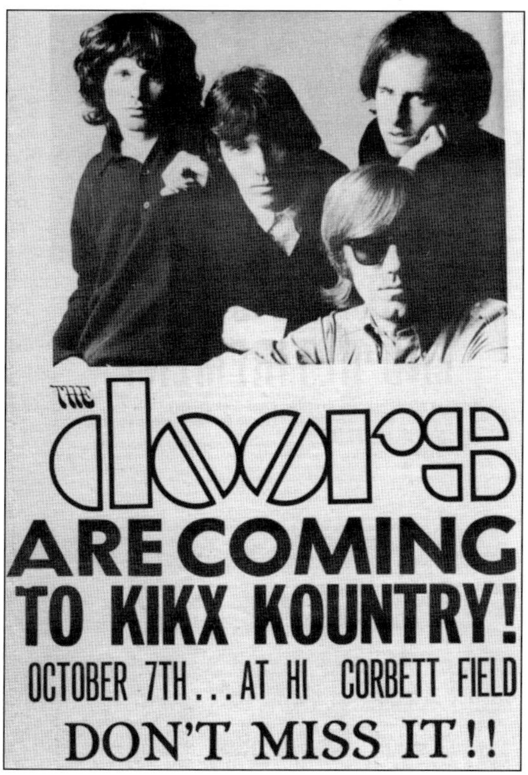

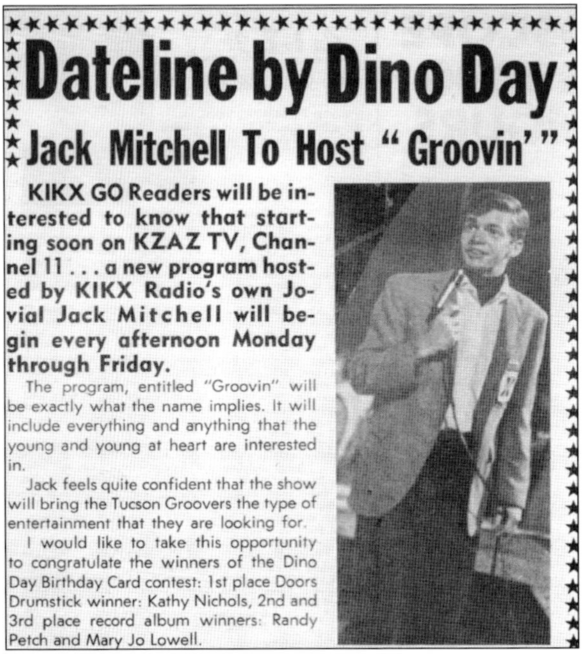

Dateline by Dino Day

Jack Mitchell To Host "Groovin'"

KIKX GO Readers will be interested to know that starting soon on KZAZ TV, Channel 11 . . . a new program hosted by KIKX Radio's own Jovial Jack Mitchell will begin every afternoon Monday through Friday.

The program, entitled "Groovin" will be exactly what the name implies. It will include everything and anything that the young and young at heart are interested in.

Jack feels quite confident that the show will bring the Tucson Groovers the type of entertainment that they are looking for.

I would like to take this opportunity to congratulate the winners of the Dino Day Birthday Card contest: 1st place Doors Drumstick winner: Kathy Nichols, 2nd and 3rd place record album winners: Randy Petch and Mary Jo Lowell.

KIKX and KZAZ-TV teamed up to organize a locally produced teen music program that aired weekdays at 4:00 p.m. The show, called *Groovin'*, was hosted by KIKX DJ Jack Mitchell. The 30-minute program featured live local bands, records, and news. In December 1967, the show was renamed *Discotheque 68*. KIKX evening DJ Peter Huntington May took over as host in January 1968. In March, the program moved to 5:30 p.m. In early May, the program left the air and was replaced with *Make Room for Daddy*, starring Danny Thomas. During the summer of 1967, when Mitchell did his regular noon program, he used various poolside sound effects, such as water splashing, in a low-key effort to portray the Sands Hotel location of the KIKX studios. This article is from the weekly *KIKX Go* magazine. (CJB.)

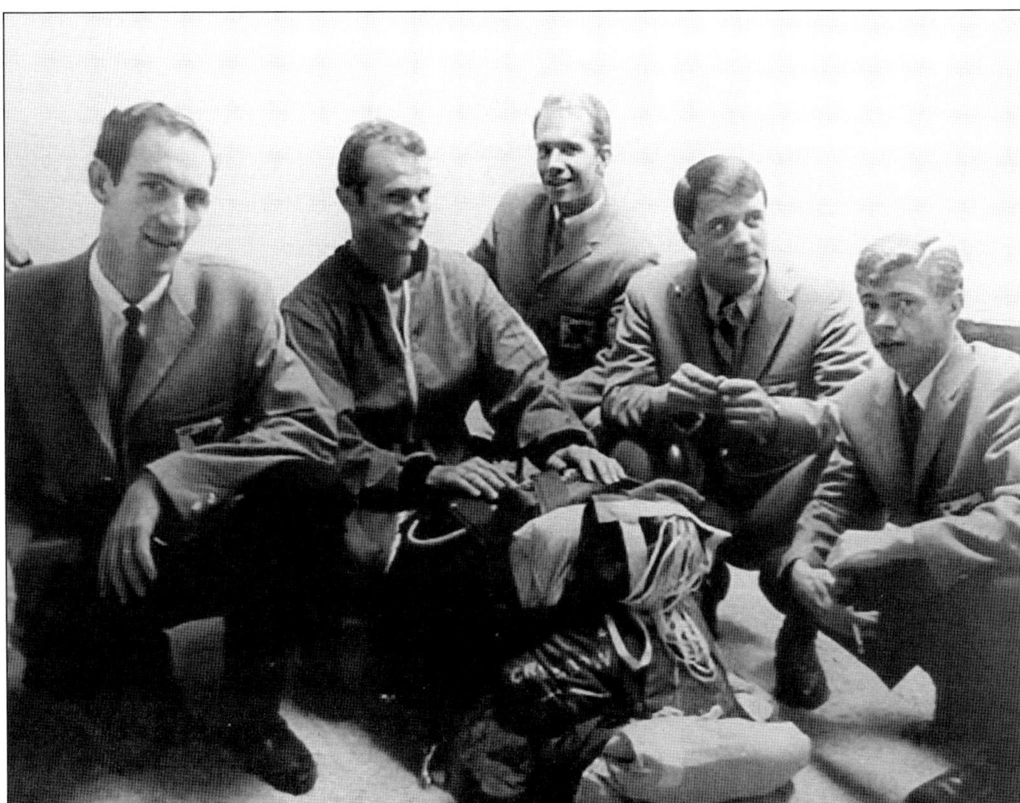

Here, four KIKX DJs pose with a skydiver at a summer 1967 promotional event. From left to right are Ted Behr, unidentified, Russ O'Hara, Jefferson K., and Jack Mitchell. (Shadoe Stevens.)

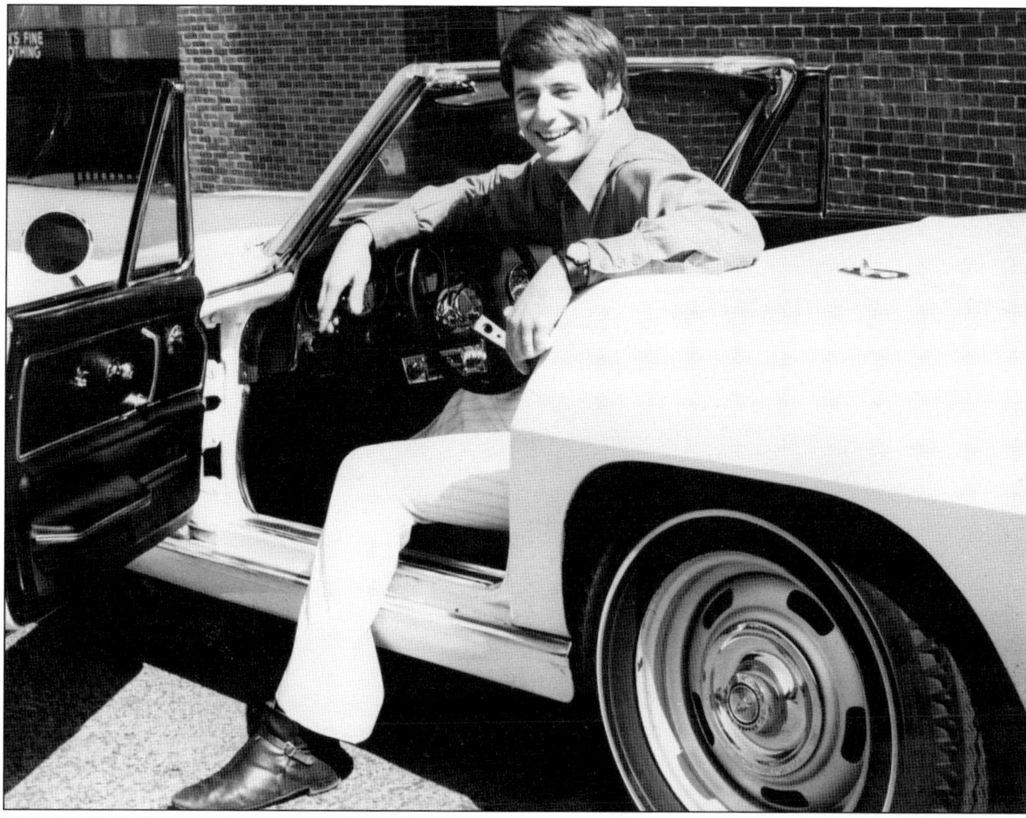

KIKX DJ "Jefferson K" previously worked at KQWB in Fargo, North Dakota. Blessed with a deep, smooth voice, he quickly became KIKX's most popular personality. Doing the late afternoon show, he soon developed a large following. In the fall of 1968, he switched to the morning program. After leaving KIKX in March 1969, he worked at Top 40 giant WRKO in Boston. Upon arriving in Boston, he adopted the name Shadoe Stevens. After WRKO, he went on to radio fame in Los Angeles at KHJ, KRLA, KMET, and KROQ. He also did a six-year stint succeeding Casey Kasem as host of the nationally syndicated radio program *American Top 40*. In addition to radio, Stevens found success doing film and television work, starring in the hit CBS television show *Dave's World* for 71 episodes from 1993 to 1997. Starting in 1998, Stevens became a regular on *Hollywood Squares*, appearing in more than 650 episodes. In the 1968 photograph above, Jefferson K is showing off his 1967 Corvette. (Above, Shadoe Stevens; right, CJB.)

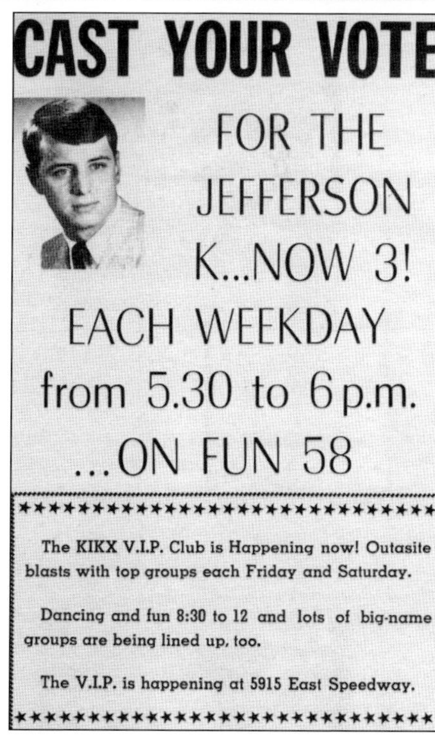

THE KIKX SIX — TOP 30 TUNEDEX

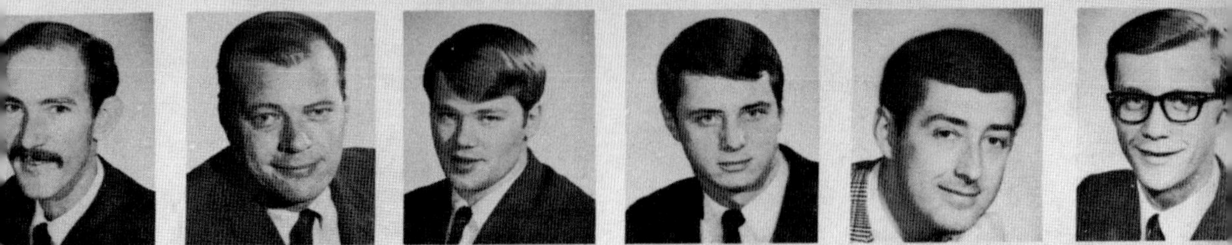

TED BEHR	DAN'L GATES	DINO DAY	JEFFERSON K	PETER H. MAY	JIM ARNOLD
6:00-9:00 a.m.	9:00-12:00 noon	12:00-4:00 p.m.	4:00-8:00 p.m.	8:00-12:00 midnight	12 Mid.-6:00 a.m

Title	Artist	Last week
LOVE IS ALL AROUND	The Troggs	1
La La Means I Love You	The Delfonics	14
Playboy	Gene & Debbe	7
Love Is Blue	Paul Mauriat	2
Summertime Blues	Blue Cheer	13
Cry Like A Baby	Boxtops	17
I Love You	People	5
Valleri/Tapioca Tundra	Monkees	16
Scarborough Fair	Simon & Garfunkel	8
Gonna Make You Love Me	Madeline Bell	11
Words	BeeGees	3
Kiss Me Goodbye	Petula Clark	15
Question of Temptation	Balloon Farm	6
Springfield Plane	Kenny O'Dell	4
Ballad of Bonnie & Clyde	Georgie Fame	10
16. Too Much Talk	P. Revere & Raiders	
17. Simon Says	1910 Fruitgum Co.	1
18. Young Girl	Union Gap	**
19. Suddenly You Love Me	Tremeloes	2
20. Since You've Been Gone	Aretha Franklin	2
21. Jennifer Juniper	Donovan	2
22. Up To My Neck	Linda Ronstadt	2
23. Rice Is Nice	Lemon Pipers	2
24. Lady Madonna/Inner Light	Beatles	2
25. In Need of a Friend	The Cowsills	2
26. We're A Winner	Impressions	**
27. There Is	The Dells	2
28. Honey	Bobby Goldsboro	**
29. L. David Sloane	Michele Lee	**
30. Call Me Lightning	The Who	**

Winning KTKT's younger demo during the first year, KIKX now wanted to widen its success. In the fall of 1968, KIKX introduced a leaner and more mature Top 40 sound. Promotions aimed at teens were scaled back and the Drake-style jingles were replaced. The weekly *Go* magazine was replaced with a stand-alone chart. A late-night "heavy music" program hosted by DJ Jim Stagg, *The Open Lid*, was added. Jim, aka Terry MacDonald, has gone on to national fame, today doing voiceover work for film and TV. KIKX seemed to be moving in the right direction when several other changes took place. In March 1969, program director Gary Palant departed for CKFH in Toronto. Jefferson K also left at this time. On May 26, 1969, newly appointed general manager Bill Crawford abruptly dropped Top 40 for an MOR format. All music aired was now tightly controlled by a computer-generated lineup. KIKX debuted an easy listening jingle package from Dynamic Productions of Dallas, Texas. Ironically, these changes took place just before the latest ratings showed KIKX finishing in the top three during all time periods. All hopes for KIKX becoming a heritage rocker in Tucson were now lost. The station would rock again in the 1970s, but the hope and magic of "Kicks 58" evaporated. The KIKX chart pictured here is from March 1968. (CJB.)

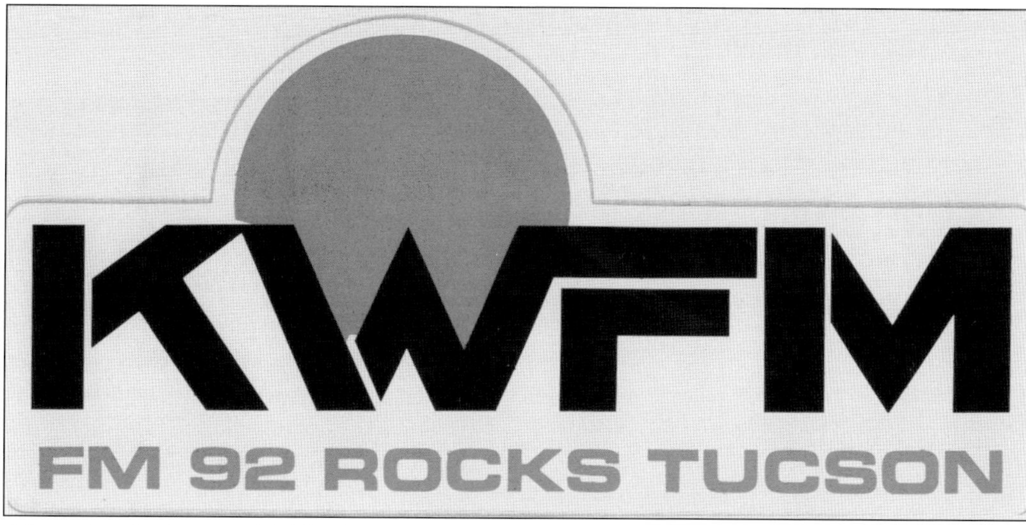

When the 1970s started, KTKT again had AM Top 40 to itself. A new type of challenger surfaced when KWFM FM 92.9 signed on in April 1970. Sister of longtime Tucson Spanish-language station KEVT AM 690, and sharing the studio facilities at 48 East Broadway, this new FM station offered Tucsonans what was then called "progressive rock" with a relaxed presentation. Catching on quickly on the UA campus, KWFM continued to grow its audience as more people and automobiles acquired FM radios. (CJB.)

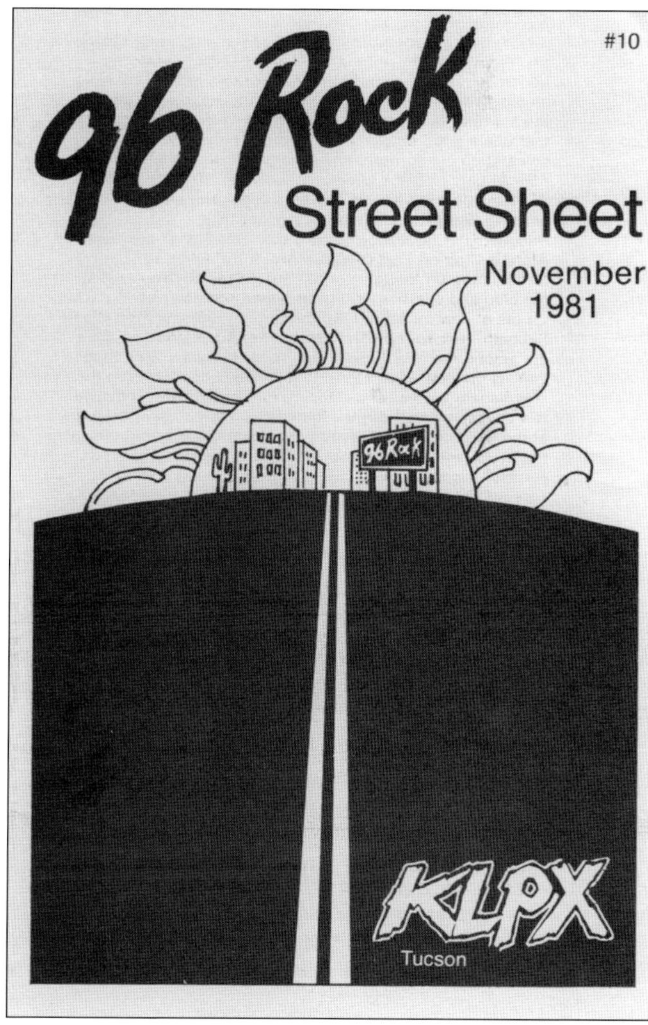

KTKT did little to directly combat FM penetration until very late in the decade. In December 1979, KTKT purchased KCEE-FM. KTKT gave it an AOR format in an effort to protect the AM ratings by further splitting the FM audience. The call letters, KTKT-FM at startup, changed to KLPX in the spring of 1981. This is a KLPX Street Sheet LP chart cover from November 1981. (CJB.)

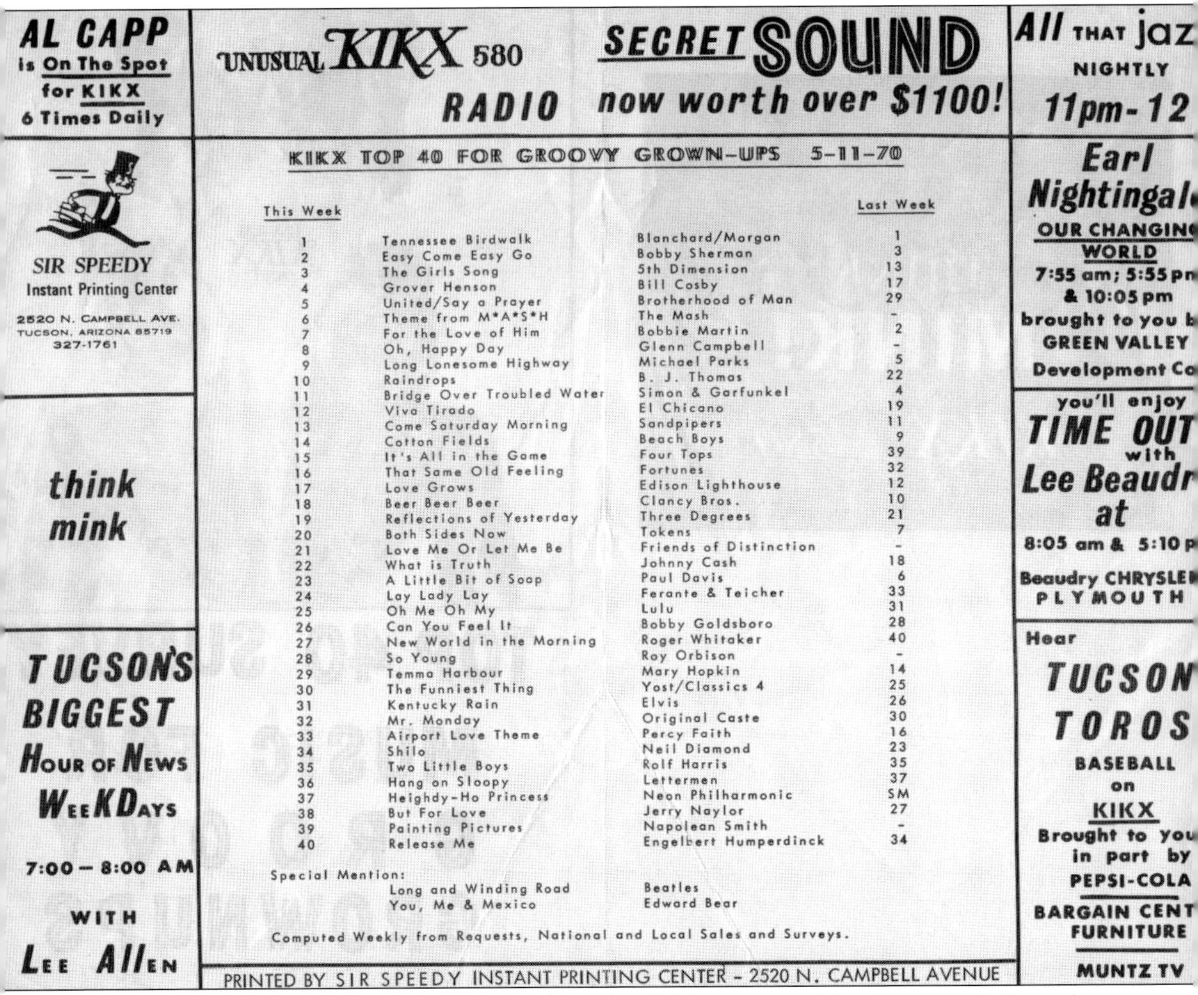

AL CAPP is On The Spot for KIKX 6 Times Daily

UNUSUAL KIKX 580 RADIO

SECRET SOUND now worth over $1100!

All THAT jaz NIGHTLY 11pm-12

SIR SPEEDY Instant Printing Center 2520 N. CAMPBELL AVE. TUCSON, ARIZONA 85719 327-1761

think mink

TUCSON'S BIGGEST HOUR OF NEWS WEEKDAYS 7:00 – 8:00 AM WITH LEE AllEN

Earl Nightingal OUR CHANGIN WORLD 7:55 am; 5:55 pn & 10:05 pm brought to you b GREEN VALLEY Development Co

you'll enjoy TIME OUT with Lee Beaudr at 8:05 am & 5:10 p Beaudry CHRYSLE PLYMOUTH

Hear TUCSON TOROS BASEBALL on KIKX Brought to you in part by PEPSI-COLA BARGAIN CENT FURNITURE MUNTZ TV

KIKX TOP 40 FOR GROOVY GROWN-UPS 5-11-70

This Week			Last Week
1	Tennessee Birdwalk	Blanchard/Morgan	1
2	Easy Come Easy Go	Bobby Sherman	3
3	The Girls Song	5th Dimension	13
4	Grover Henson	Bill Cosby	17
5	United/Say a Prayer	Brotherhood of Man	29
6	Theme from M*A*S*H	The Mash	-
7	For the Love of Him	Bobbie Martin	2
8	Oh, Happy Day	Glenn Campbell	-
9	Long Lonesome Highway	Michael Parks	5
10	Raindrops	B. J. Thomas	22
11	Bridge Over Troubled Water	Simon & Garfunkel	4
12	Viva Tirado	El Chicano	19
13	Come Saturday Morning	Sandpipers	11
14	Cotton Fields	Beach Boys	9
15	It's All in the Game	Four Tops	39
16	That Same Old Feeling	Fortunes	32
17	Love Grows	Edison Lighthouse	12
18	Beer Beer Beer	Clancy Bros.	10
19	Reflections of Yesterday	Three Degrees	21
20	Both Sides Now	Tokens	7
21	Love Me Or Let Me Be	Friends of Distinction	-
22	What is Truth	Johnny Cash	18
23	A Little Bit of Soap	Paul Davis	6
24	Lay Lady Lay	Ferante & Teicher	33
25	Oh Me Oh My	Lulu	31
26	Can You Feel It	Bobby Goldsboro	28
27	New World in the Morning	Roger Whitaker	40
28	So Young	Roy Orbison	-
29	Temma Harbour	Mary Hopkin	14
30	The Funniest Thing	Yost/Classics 4	25
31	Kentucky Rain	Elvis	26
32	Mr. Monday	Original Caste	30
33	Airport Love Theme	Percy Faith	16
34	Shilo	Neil Diamond	23
35	Two Little Boys	Rolf Harris	35
36	Hang on Sloopy	Lettermen	37
37	Heighdy-Ho Princess	Neon Philharmonic	SM
38	But For Love	Jerry Naylor	27
39	Painting Pictures	Napolean Smith	-
40	Release Me	Engelbert Humperdinck	34

Special Mention:

Long and Winding Road	Beatles
You, Me & Mexico	Edward Bear

Computed Weekly from Requests, National and Local Sales and Surveys.

PRINTED BY SIR SPEEDY INSTANT PRINTING CENTER – 2520 N. CAMPBELL AVENUE

On the AM side, KTKT faced more rivals as the 1970s went on. The 1970 KIKX chart shown here indicates a cluttered broadcast day. KIKX gave up on the groovy grown-ups MOR format in late 1971 to rock one more time. The station returned to a Top 40 format with a good-sounding, mature presentation. By 1974, KIKX was enjoying some ratings success against KTKT. But any success was dampened by a promotional event and contest that went wrong. A missing-person alert for new KIKX morning DJ Gary Craig was announced over the air. Listeners would win prizes for guessing his location. KIKX also referred to the stunt as a kidnapping during newscasts on January 19. Concerned local residents flooded the Tucson police with questions regarding the matter. While aimed at simply promoting the new KIKX personality, the contest backfired. For a couple of days, KIKX staff failed to tell authorities and the other media the kidnapping was a stunt. After discovering the truth, members of the media and law enforcement became incensed. (CJB.)

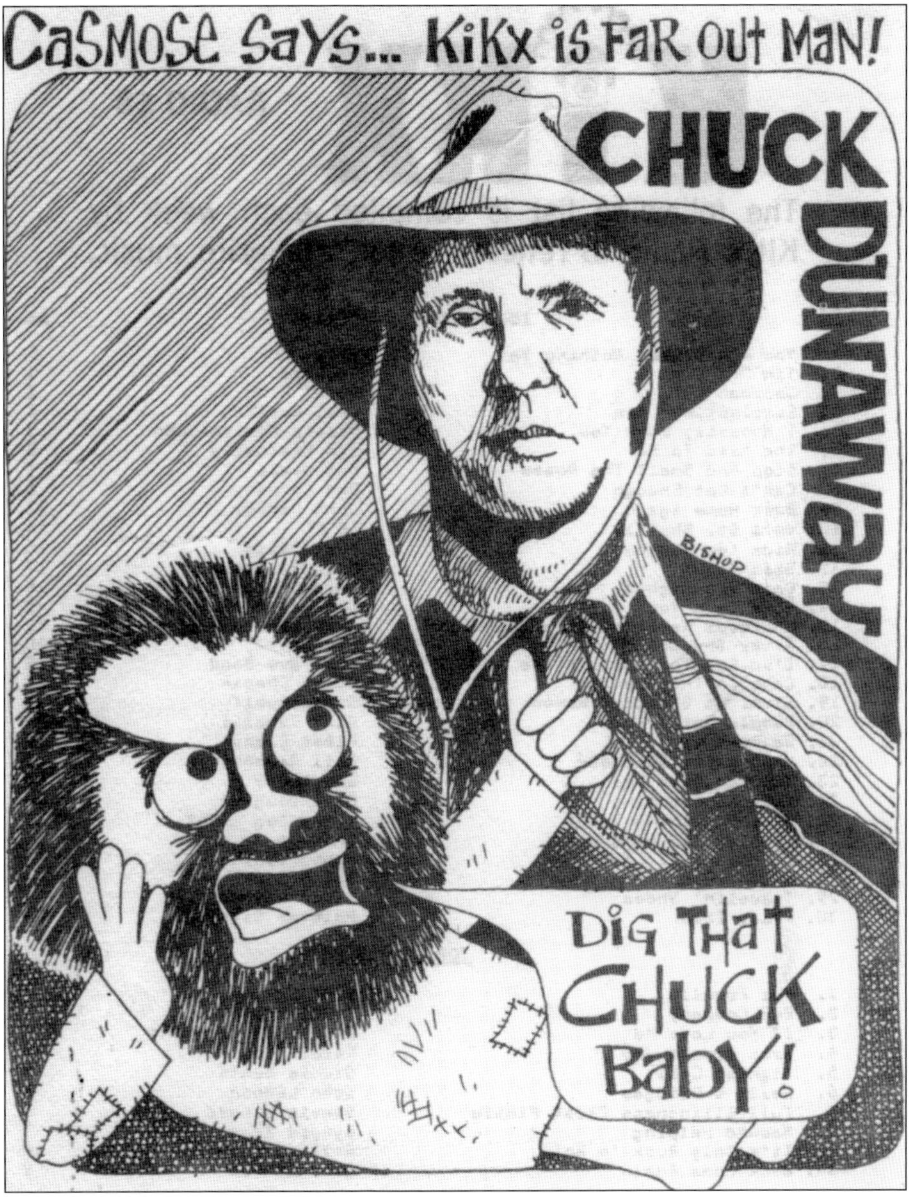

In the wake of the Gary Craig stunt, a newspaper reporter with a friend at the FCC in Washington, DC, made a phone call. Tucson chief of police William Gilkinson opened an investigation and complained in writing to the FCC about the deception. Prompted by these complaints that KIKX had not acted in the public interest, the FCC investigated. By June 1974, KIKX, under a new general manager, began damage control. After formal hearings over the next 18 months, the FCC decided to deny the station's license renewal due in October 1976. KIKX contested, using every appeal. A change from Top 40 to country was made in October 1977. DJ Bob Jones played the last rock record, "God Only Knows" by the Beach Boys, just before the midnight format change. After several failed appeals to the FCC, KIKX finally gave up the legal battle and went dark at midnight on Saturday, July 17, 1982. KIKX played "The Last Cowboy Song" by Ed Bruce just before going dark some 15 years after signing on with such great promise. This 1974 chart cover shows DJ Chuck Dunaway and mascot Casmose. (CJB.)

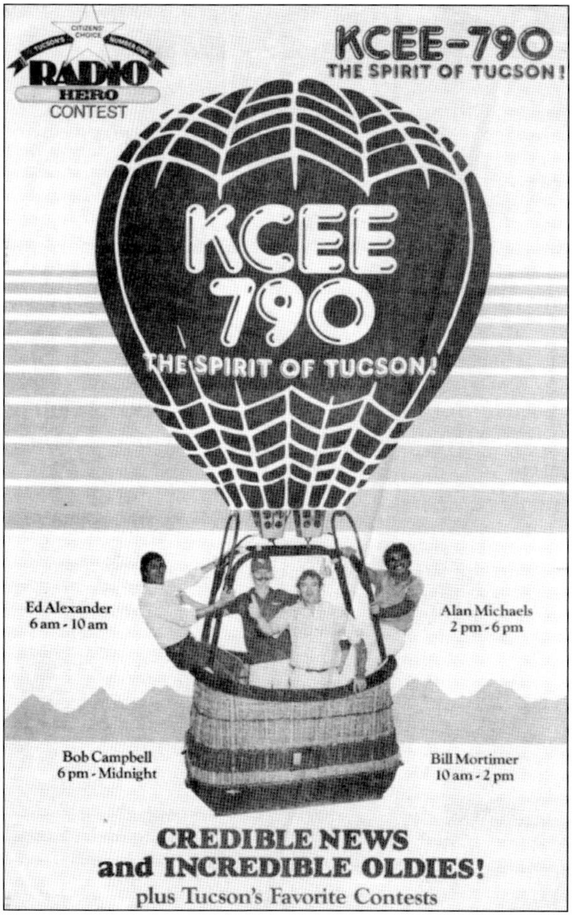

Feeling the growth of FM, Tucson AM stations kept evolving in the 1980s. KHYT owner Bob Scholz hired Rich "Brother" Robbin from San Diego in late 1979 to be the program director. Since inception, KHYT aired several types of pop music formats. Robbin gave the station a great-sounding, fast-paced Top 40 format, heavy on dance hits. KHYT gained 24-hour authority during early January 1980. In the mid-1980s, Robbin undertook another Top 40 effort as owner of AM 1450, making the station KKPR, "Power Hits." After adding FM 92.1, Robbin re-branded the FM station as KFXX "The Fox." Both featured Top 40 music with a heavy dose of dance hits. After a couple of years, however, both stations were sold and dropped Top 40 as a format. KCEE AM 790 (left) shifted from light pop to full-time rock 'n' roll oldies in the early 1980s, employing popular local DJs. Despite all of these shuffles on the AM band a new Top 40 leader emerged, and on FM, KRQQ (below). (Left, Bill Mortimer; below, RJ.)

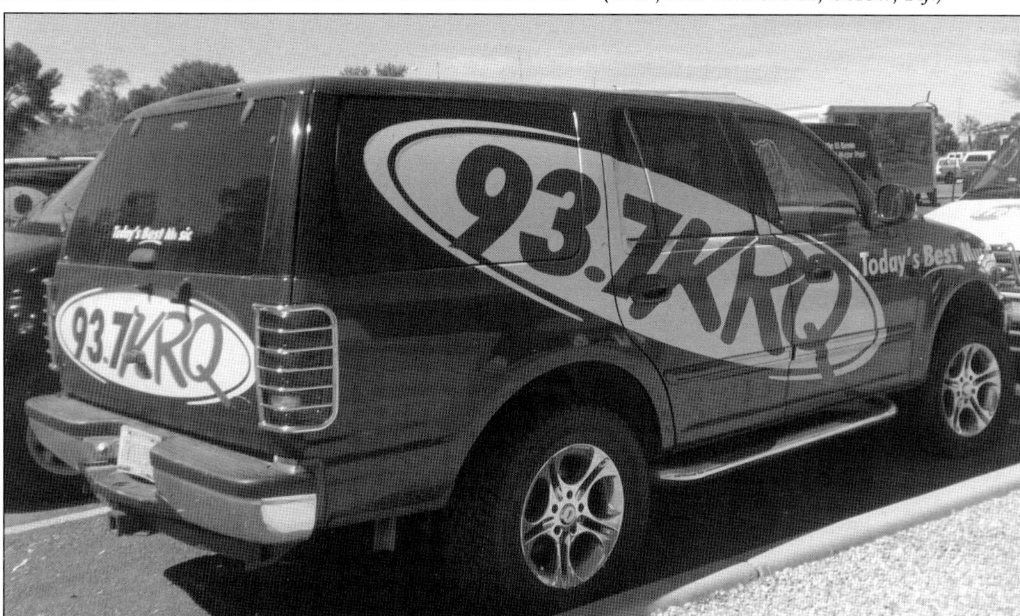

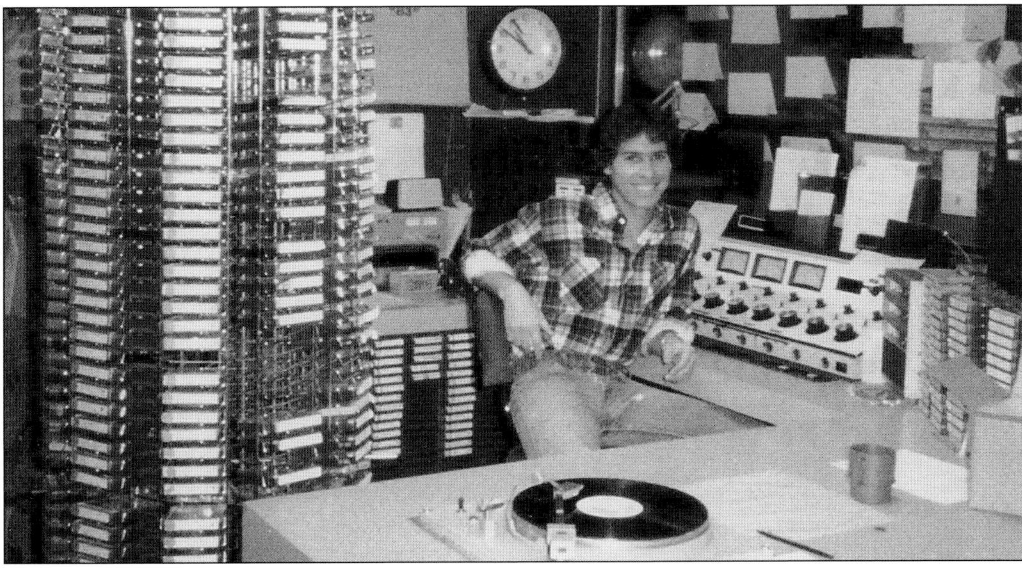

KRQQ began life as KXEW-FM 93.7 in 1971, allowing the daytime AM station KXEW 1600 to be heard 24 hours a day with programming in Spanish. In 1977, the FM station was purchased by Bill Phalen, then the owner of KHOS, who managed to put the station on the air with a new transmitter. Shown here are two of the first KRQQ disc jockeys, Jeff Davis (above) and Kent Lacy (below), who did afternoons. In the first year, according to Davis, "there were pre-recorded breaks on tape until the station went 'live' in March of 1978 from the new La Placita Village studio." The studio is seen in these photographs. Davis added, "yes, complete with a turntable!" During these times, 93.7 K-R-Q, as it has always been known on the air, described itself as a Top 40, popular, or contemporary music station. It never attempted to sound heavy, underground, progressive, or free form. This approach gave the last group of KTKT listeners an FM station they could tolerate. Phalen changed KHOS from country to KMGX, also a Top 40 contemporary music station, to further fragment KTKT's audience on the AM side. (Jeff Davis.)

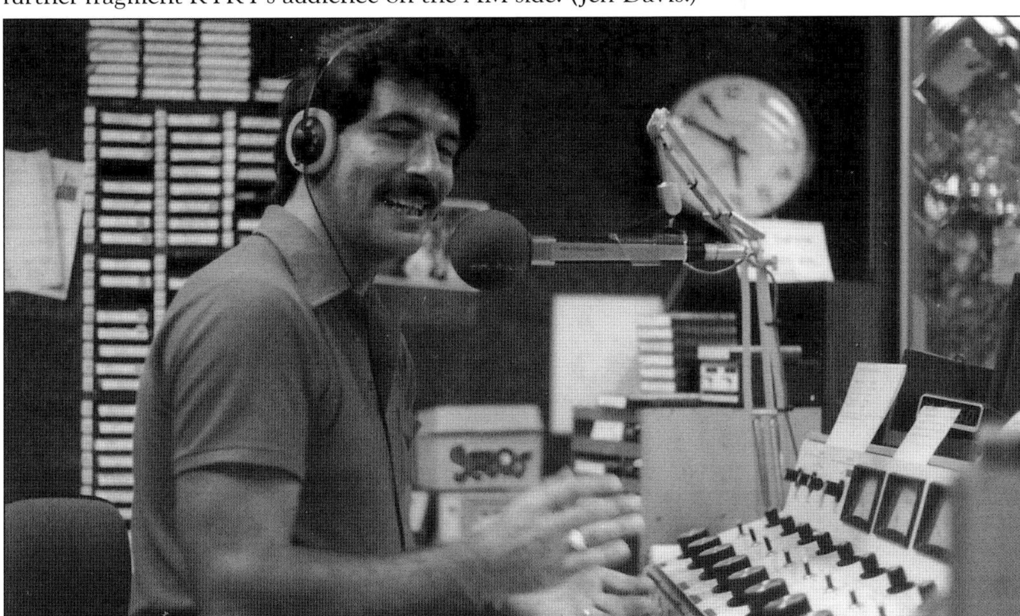

Tucson's FM (Favorite Music)

Last Week	This Week	Title	Artist
2	1	Last Train To London	Elo
1	2	Rock With You	Michael Jackson
5	3	Coward Of The County	Kenny Rogers
3	4	Sara	Fleetwood Mac
6	5	Romeo's Tune	Steve Forbert
8	6	Why Me	Styx
9	7	Crazy Little Thing Called Love	Queen
4	8	Ladies Night	Kool & The Gang
10	9	You Know That I Love You	Santana
16	10	Longer	Dan Fogelberg
7	11	Don't Do Me Like That	Top Petty
15	12	Wait For Me	Hall & Oates
11	13	Dammed If I Do	Alan Parsons
17	14	Third Time Lucky	Foghat
19	15	Savannah Nights	Tom Johnston
18	16	I Do Believe In You	Pages
21	17	"99"	Toto
20	18	Heart Breaker	Pat Benatar
27	19	I'd Rather Leave While I'm In Love	Rita Coolidge
13	20	Deja Vu	Dionne Warwick
24	21	Yes, I'M Ready	Teri Desario
26	22	An American Dream	Dirt Band
12	23	We Don't Talk Anymore	Cliff Richard
25	24	I Want To Be Your Lover	Prince
14	25	Escape	Rupert Holmes
29	26	Lonely Eyes	Robert John
31	27	Day Dream Believer	Ann Murray
30	28	Dirty Water	Inmates
22	29	Head Games	Foreigner
35	30	Remember (Walking In The Rain)	Aerosmith
40	31	Working My Way Back To You	Spinners
ADD	32	Refugee	Tom Petty & The Heartbreakers
36	33	Lost Her In The Sun	John Stewart
ADD	34	Another Brick In The Wall	Pink Floyd
38	35	I'm Alive	Gamma
D	36	When I Wanted You	Barry Manilow
ADD	37	September Morn	Neil Diamond
33	38	I Want You Tonight	Pablo Cruise
23	39	Fool In The Rain	Led Zepplin
ADD	40	Don't Let Go	Issac Hayes

This list has been compiled by the KRQ Research Dept. Numbers do not necessarily indicate amount of airplay.

KRQ 94 Rock FM Concert List

January 21st	Muddy Waters	Dooley's/Tucson
January 22nd	Tom Petty/Fabulous Poodles	Symphony Hall/Phoenix
January 24th	Pat Benatar	Dooley's/Tempe
January 26th	Tom Petty/Fabulous Poodles	UofA Main Auditorium
January 26th	Willy Nelson/Kristofferson	Phoenix Coliseum
January 29th	Willy Nelson/Kristofferson	TCC/Arena
January 31st & February 1st & 2nd	Neil Sedaka	Symphony Hall/Phoenix
February 4th	Angela Bosill Jazz Show	Dooley's/Tucson
April 15th	Harry Belafonte	ASU Activity Ctr./Tempe
March 1st	Chuck Mangione	UofA Main Auditorium

Special Features

Monday thru Friday
1:55 & 10:55 P.M.
KRQ Mini Concerts

Friday 12 Midnight
Dr. Demento

Saturday 12 Midnight
KRQ Amateur Hour

Monday thru Friday
at 7:55
KRQ Rock Block

Saturday Nite —
Beginning at One in the morning
KRQ Classic Album
All-night Album Replay

Sunday Nite 8 PM —
Great American Radio Show
Top 20 Album Trax of the Week

Sunday Nite 10 PM
Alien Worlds

KRQ Line Up

6 AM-10 AM - Dan McCoy
10 AM-3 PM — Jim Gillie
3 PM-7 PM — Steve Goddard
7 PM-12 Mid — Steve Rivers
12 Mid-6 AM — Jeff Davis
Weekends — John Stevens
Toby Bell

Numbers to Remember

KRQ Concert Line
882-9494
KRQ Request Line
888-5000
TCC Box Office
791-4266

Tucson's Best Rock KRQ FM

Jeff Davis says that, for a time under program director Guy Zapoleon, "he had KRQ playing The Zombies, old Rolling Stones songs, The Seeds, Mitch Ryder and the Detroit Wheels, all kinds of oldies rock. Yet in between we were playing The Cars, Blondie, and post disco stuff. It really didn't work and we tanked. He pulled it together, though, and he ended up being a finalist for Program Director of the Year. Because of him we were also the first station in the country to play the Stray Cats and one of the first songs by the Go-Go's. Guy knew his stuff." Zapoleon had previously been music director at oldies station KRTH in Los Angeles. It took KRQ until 1980 to finally beat KTKT. One of the early voices on KRQ was Ed Alexander, who remained throughout his career in Tucson broadcasting. Davis says, "He sure could have made a name for himself had he moved on to bigger markets. What a voice!" From 1984 to 1987, KRQ had television shows on MCaw cable television because of the small number of homes wired for cable. Few people could see MTV music videos except on these programs. (CJB.)

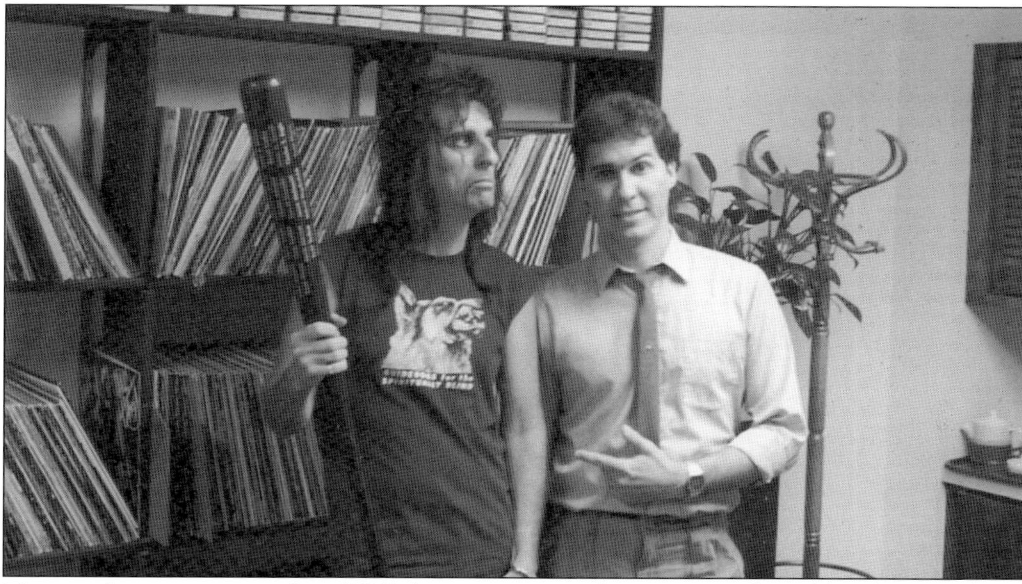

Many rock artists appeared on KRQ for interviews and to guest on the cable television shows. Here, rock superstar Alice Cooper exchanges words with KRQ's Jeff Davis in 1986. Davis spent a few months in 1985 in England, doing "Pirate Radio." He is a member of the Pirate Radio Hall of Fame. Davis also did two years on Phoenix radio. In 1989, the stations moved to 4400 East Broadway. Among the KRQ jocks of the 1980s were Terry Daniels, Kelly Norris, Jim Miller, Mike Elliott, Charlie Rabbit, and Gary Jackson. When Davis left KRQ in 2000, he established a career in voiceover, narration, and audio production. (Jeff Davis.)

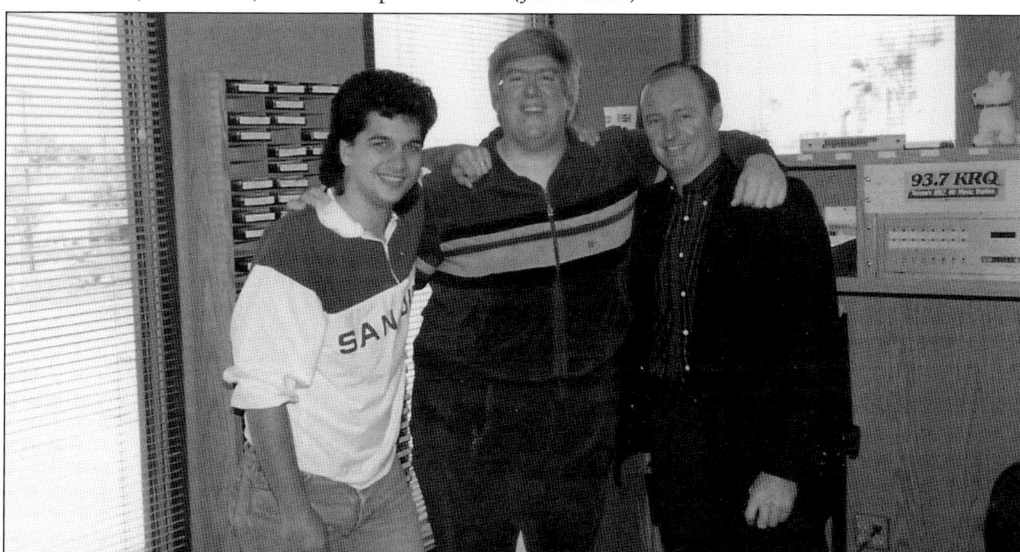

Some nationally known radio people worked at KRQ through the years. Clarke Ingram (center), the 1988–1989 program director, is seen here with air personalities Mike Elliott (left) and Clancy Woods. Another notable who worked with Elliott and Betsy Bruce in 1992–1994 was future television star Jimmy Kimmel, whose time at KRQ was probably his first foray as a daily on-air talent. Davis says, "He was always fighting with program director Mike Todd about what he could get away with on the air." (R Dub!.)

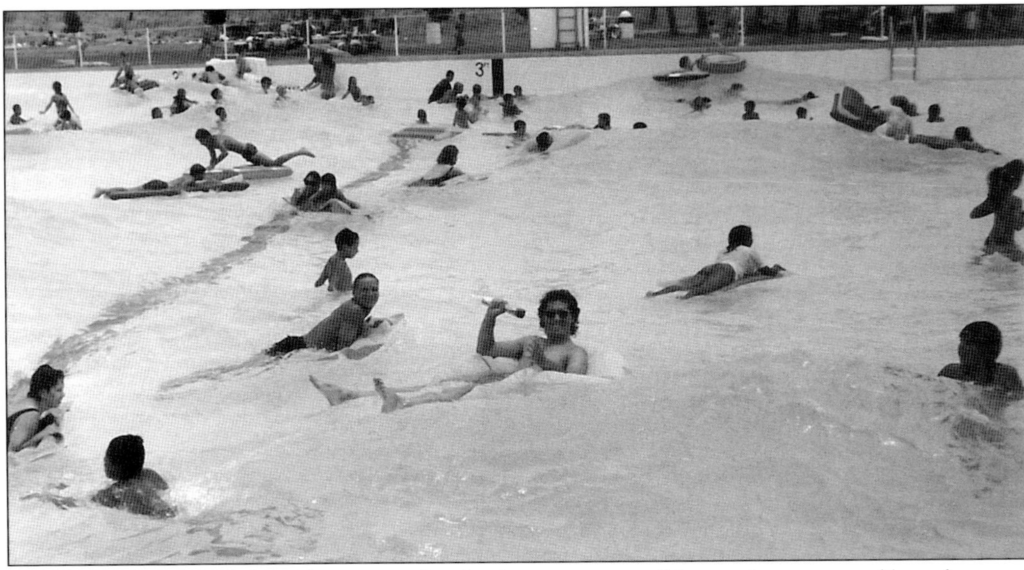

Promotions of all kinds have been a staple of radio stations, and Tucson has seen and heard some of the best. Each radio station made itself and its personalities "visible" to the audience, who flocked to see what was going on and perhaps to win prizes, ranging from records to concert tickets to cruises. In the above photograph, Jeff Davis is in the pool at the Breakers Water Park. He did his midday show there each week on "Wet & Wild Wednesday" in the summer for two years. "I would wrap the microphone in a Ziploc bag and do some breaks from [that] inner tube in the wave pool. Surprisingly our engineer never stopped me from doing it. That made for some good radio fun." When Davis left the program, the on-air "Wet & Wild" promotion and his "Breaker babes" were taken over by Roger Scott, KRQ's music director. In the below photograph, Scott interviews one of the contestants in the "Slide-a-Thon." (Above, Jeff Davis; below, R Dub!.)

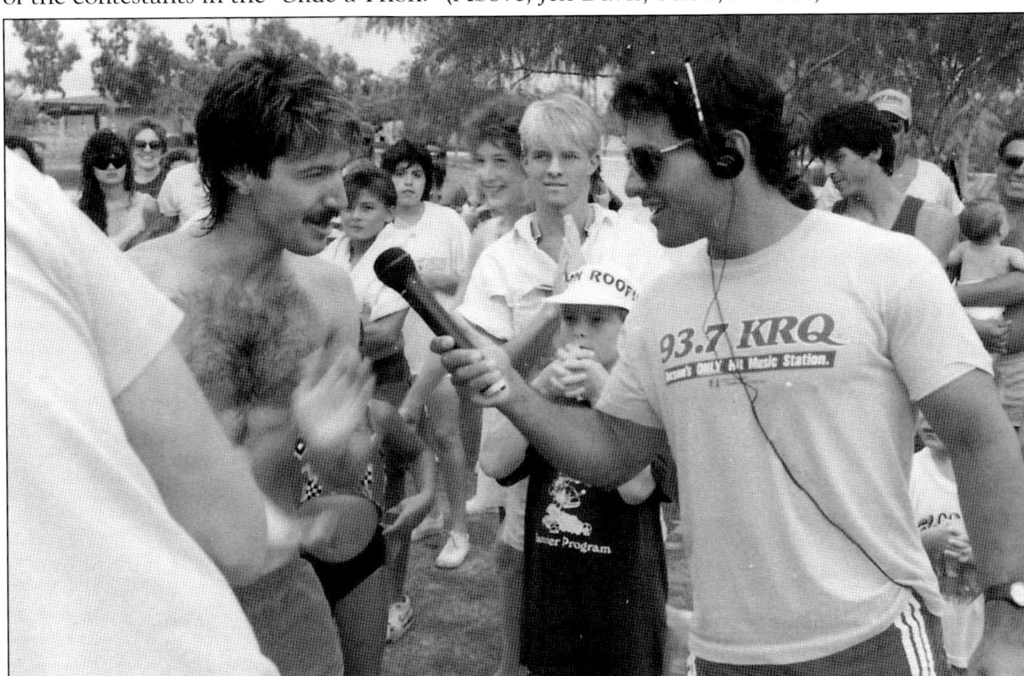

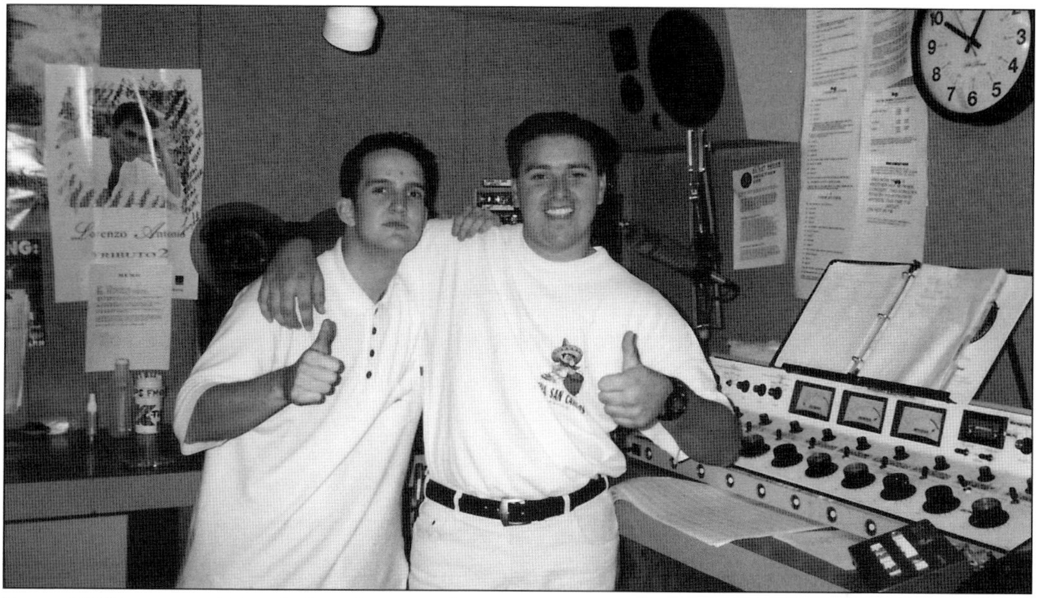

Some KRQ personalities defected to another station around 1990 and whipped up the audience with Power 1490, a rapidly growing music format of "hip-hop and rhythm & blues." For almost five years, the low-power AM station battled the FM giant with promotions and teasing while developing a dedicated audience of mostly young people. KRQ and its Top 40/contemporary hits format acquired some FM competition in the late 1980s, when formerly all-Spanish KOPO 98.3 changed its call letters to KOHT, calling itself HOT 98.3 FM. It was owned by Los Angeles radio's Art Leboe. Air personality "R Dub!" was on Power 1490 in 1994–1995. Above, he and fellow jock Kid Corona (right) are at the KOHT controls in 1996. Below, R Dub! interviews rap music artist and New York radio personality and actress Angie Martinez in 2003. (R Dub!.)

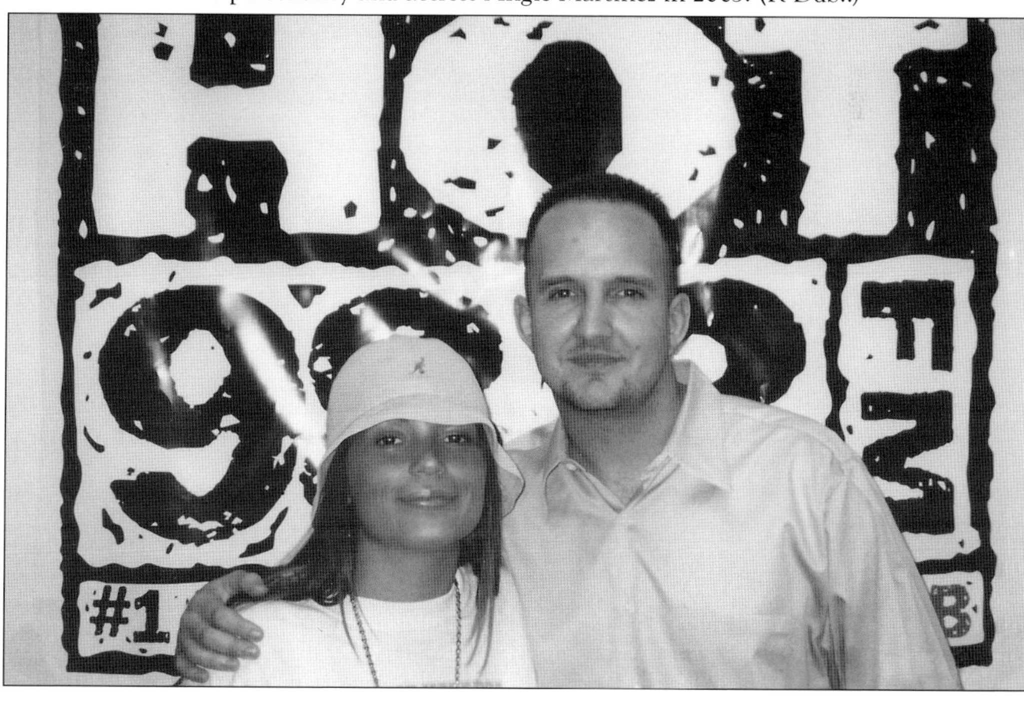

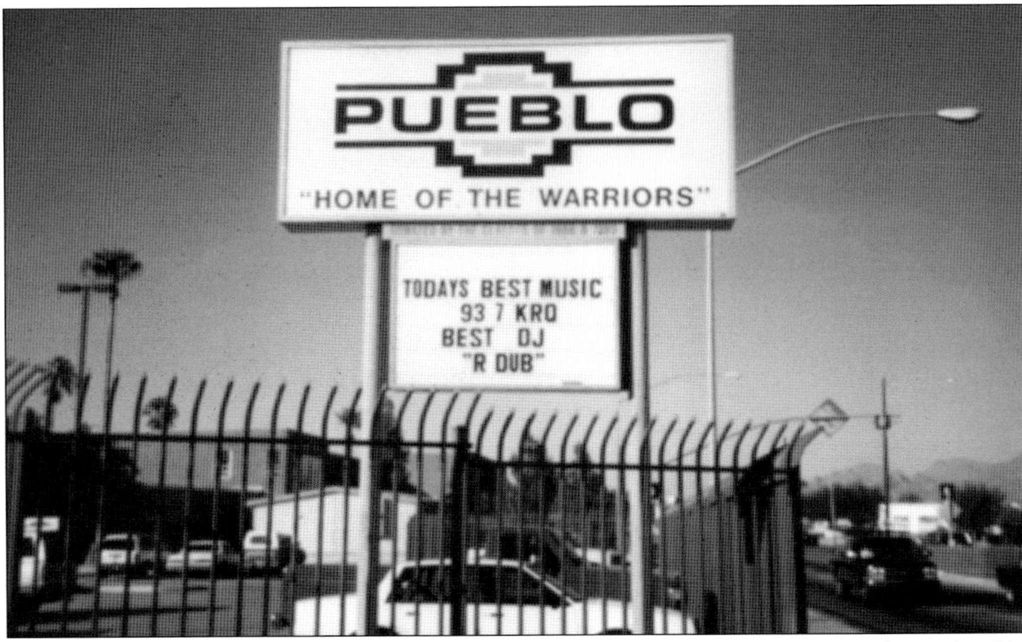

Among its promotions, KRQ held a contest for a high school to win a concert at their school by the vocal group Voices of Theory. The sign in the above photograph shows a school's participation. Students competed in nightly games to score points for their school. Below, Gary Coleman came to Tucson for a year to cohost the morning show on KRQ with Mojo and Betsy. Coleman was the diminutive actor on the television show *Diff'rent Strokes*. In the below photograph, he poses with KRQ's R Dub! outside the Gotham/New West nightclub. Mojo Morales and Betsy Bruce were big favorites of Tucson morning listeners, who tuned in for their contests and features. Morales moved to Detroit in 2000. His syndicated program returned to Tucson on KOHT and then KMIY a few years later. (R Dub!.)

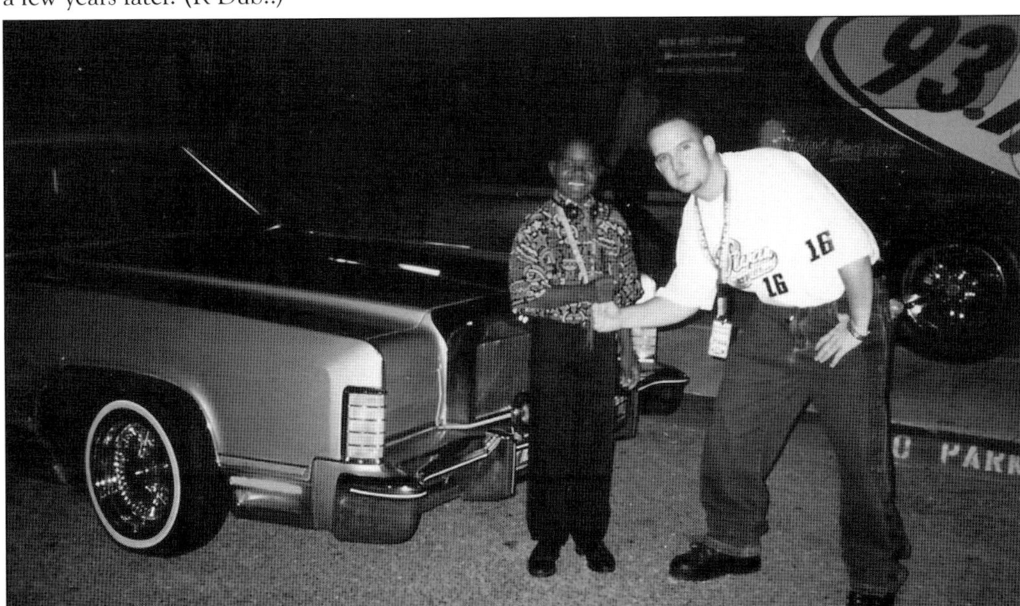

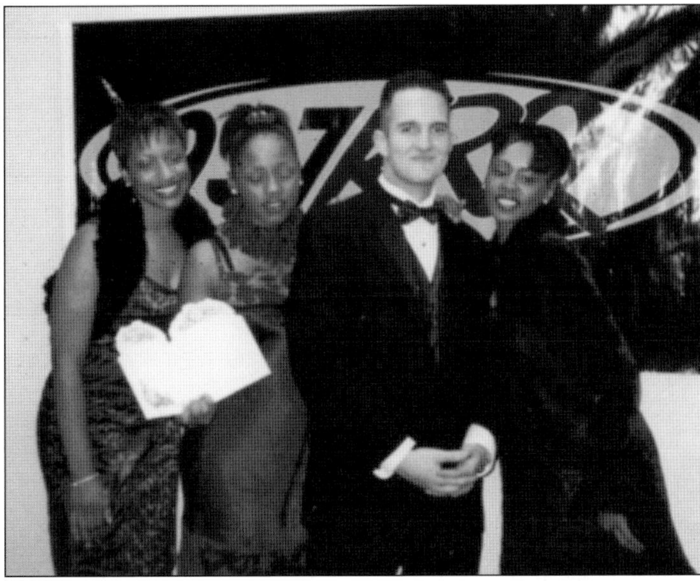

In 1999, RDub! celebrated the first annual Sunday night "Slow Jams LIVE" event at The New West. The guest star one night was the popular group Divine, shown posing with R Dub!. His four-hour program, *Sunday Nite Slow Jams,* is a fixture on Tucson radio. It began on KRQ, and in 2003, it moved to its new home on "Rhythmic Contemporary Hit Radio" KOHT. The program is also syndicated on 50 radio stations around the world. (R Dub!.)

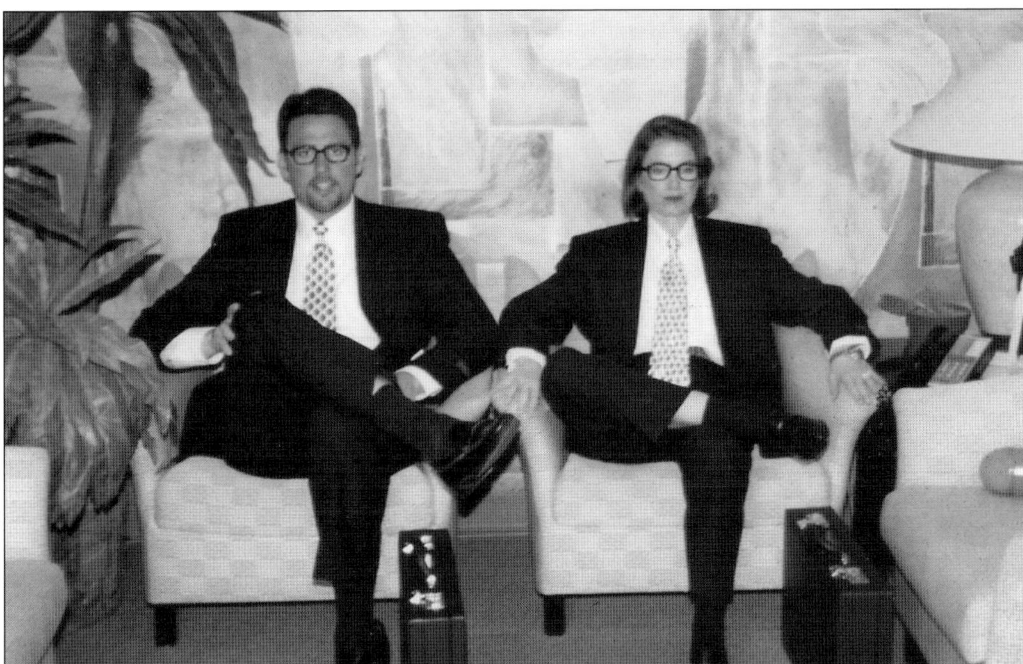

In the late 1990s, KRQ's morning team, "Mojo and Betsy," produced memorable television promotions humorously calling attention to their program. They dressed up as different people throughout Tucson, including mailmen, motorcycle cops, jet-fighter pilots, hippies, doctors, and, in this photograph, folks waiting to see the doctor. The campaign aired on Tucson television channels. Mojo says, "It was a great campaign, and to this day I swear it helped us overtake KIIM in the morning." (R Dub!.)

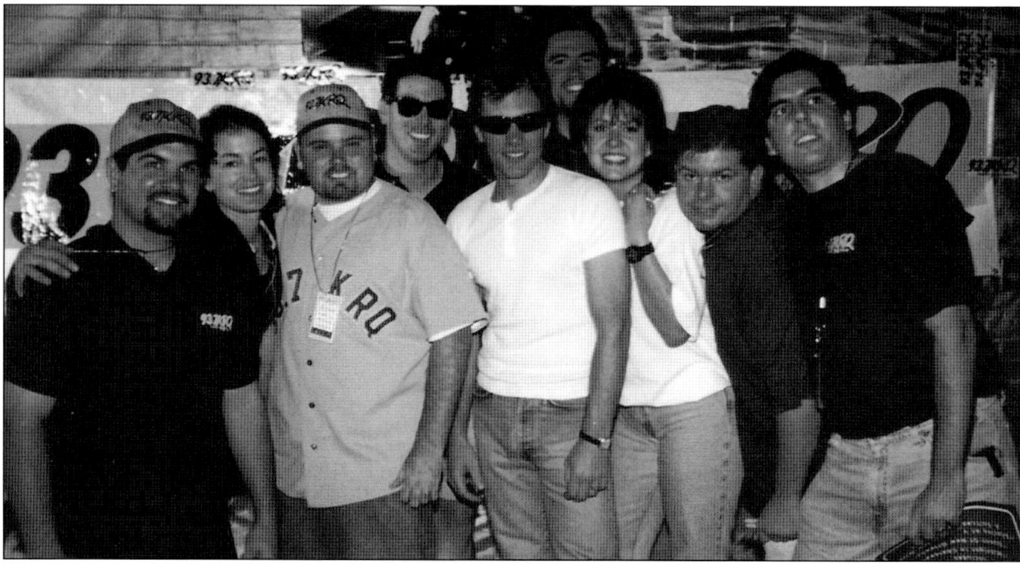

In the above photograph, Jon Bon Jovi (center) is surrounded by the KRQ air personality staff, including, from left to right, unidentified, Betsy Bruce (morning co-host), Mark Medina (nights), Adam Smasher (afternoons), Bon Jovi, Mojo Morales (morning co-host), Shannon Black (middays), Omelet (morning show producer), and Dino. This was supposed to be a "secret concert" with Bon Jovi. But, as R Dub! says, a competitor station had other plans. "KRQ started the contest about a month out—you could win your way into this exclusive, secret show, the artist and location unknown. We at Power 97.5 KSJM caught wind of who the show was with and where and ran spoiler promos on the air promoting the show at The Outback, including promos with a Jon Bon Jovi impersonator saying, 'Hey Tucson! Thanks to Power 97.5 for bringing me to town for KRQ's secret show. See you there.' We even showed up and broadcast live from across the street the day of the show!" Such was the spirit of competition between radio stations in those days. (R Dub!.)

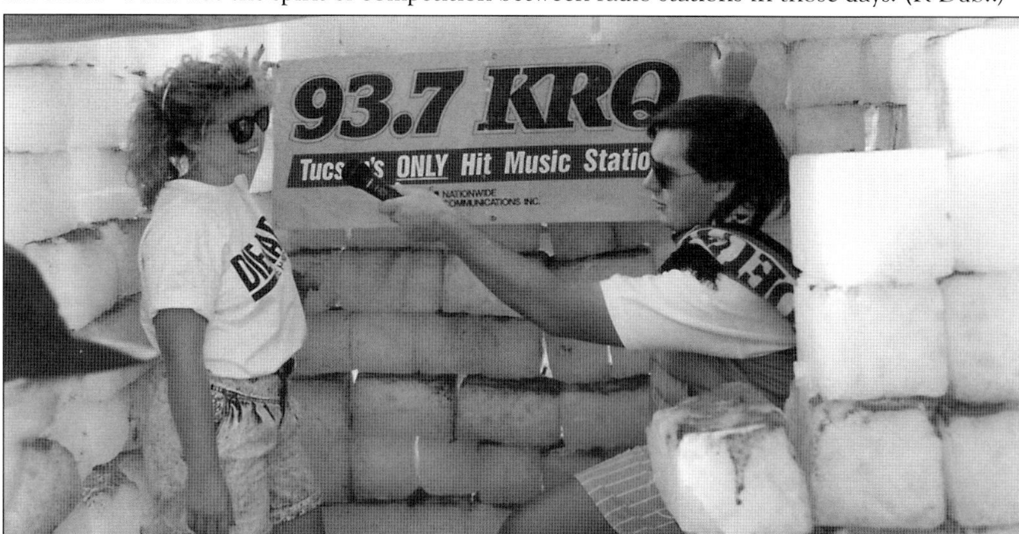

In another KRQ summer promotion, personality Bruce St. James (right) was surrounded by blocks of ice at Park Mall. St. James had bolted to Power 1490 as its program director in 1990. Radio people often move around, even in the same city. (R Dub!.)

Four

Country's Comin'
Some Hillbilly, Some Western, but All Country

Country music was rooted in the folk music of the South and Appalachia. The earliest recordings were of individual artists and groups playing songs of their heritage, the first recognized "session" being that of the Carter family in east Tennessee. Artists became more recognized as they appeared on the *Grand Ole Opry*, a live Saturday-night radio show from WSM, Nashville. The *Opry* had stars like Roy Acuff, Hank Williams, Ernest Tubb, Red Foley, and comedian Minnie Pearl. Tucson listeners first heard the *Opry* in a half-hour show on NBC radio, sponsored by Prince Albert "smokin' tobacco." KVOA had a daily program of "western" music and sponsored Saturday-night barn dances at the Pioneer Hotel, complete with live performances from Larry and the Sunset Riders. Country music records were hard to find outside the South, but they gradually became available and popular into the 1950s. KCNA announcer Al Morris left Tucson for a job in Atlanta, and when he returned, he brought back the local music he had heard there. He talked KCNA into letting him play those records on a new program called *Arizona Hayride*, which aired from 8:15 to 9:00 each morning. It was a hit. When Bob McKeehan took over the *Hayride*, the show was on all afternoon.

In 1957, KMOP 1330 came on the air, playing only country. It was soon No. 2, behind the Top 40 station, KTKT. When Jim Slone came to Tucson in 1963 to program a new station, KHOS on 940, it swept up the country audience with its continuous, 24-hour playlist. Slone took over KCUB 1290 in 1972 and, soon, it was one of the top-rated stations in the United States as listeners responded not only to the music but to its highly identifiable on-air personalities and Jim's genius for promotion. When FM radio became dominant in the 1980s, Slone acquired 99.5, and KIIM-FM was born. Although the Slone family sold its stations and Jim retired, KIIM continued to be a market leader with a franchise of gold. The artists are different, the sound is more modern, but the country music audience remains loyal to its current stars and heritage.

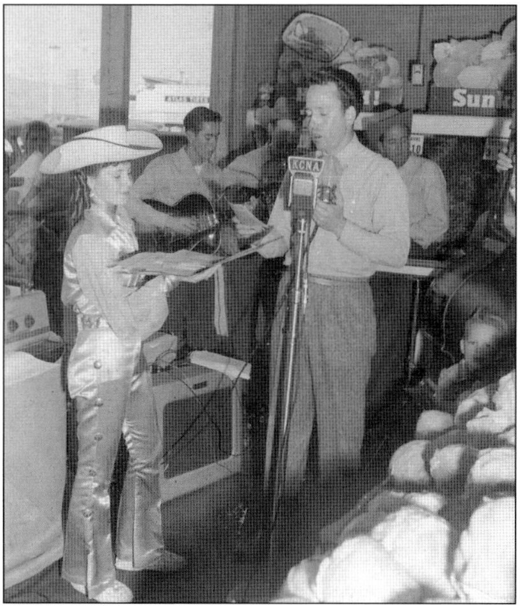

Coming from WNOX, Knoxville, Bob McKeehan spent 15 years as a Tucson country music host, first with KCNA's *Arizona Hayride* and then on KMOP, KAIR, and KHOS. Calling himself the "West Virginia–Kentucky hillbilly boy," McKeehan had an easy manner that made him welcome to listeners to KCNA's 5,000-watt signal all over southern Arizona. Here, he is doing a live broadcast from Goodman's Market at Twenty-second Street and Cherry Avenue in 1952, with his band and a guest, future singing star LaVerne Davis. When McKeehan left Tucson, it was to program religious radio stations in California. He now owns a station near Knoxville, Tennessee. (Dave Lawrence.)

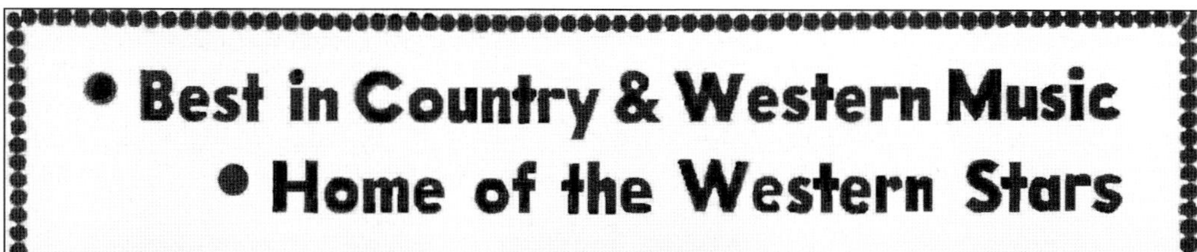

KAIR 1490 briefly tried to compete in the country music market. It hired Bob McKeehan when KCNA gave up on the *Hayride* for an easy listening, pop format. McKeehan had performed locally with the band he formed, playing weekends at the Tucson Sports Center and for special events in the 1950s. KAIR's live program, showcased on this advertisement, featured his band. The KAIR experiment with the country format was short, as the station decided to go back to rock 'n' roll in an attempt to compete for listeners. (CJB.)

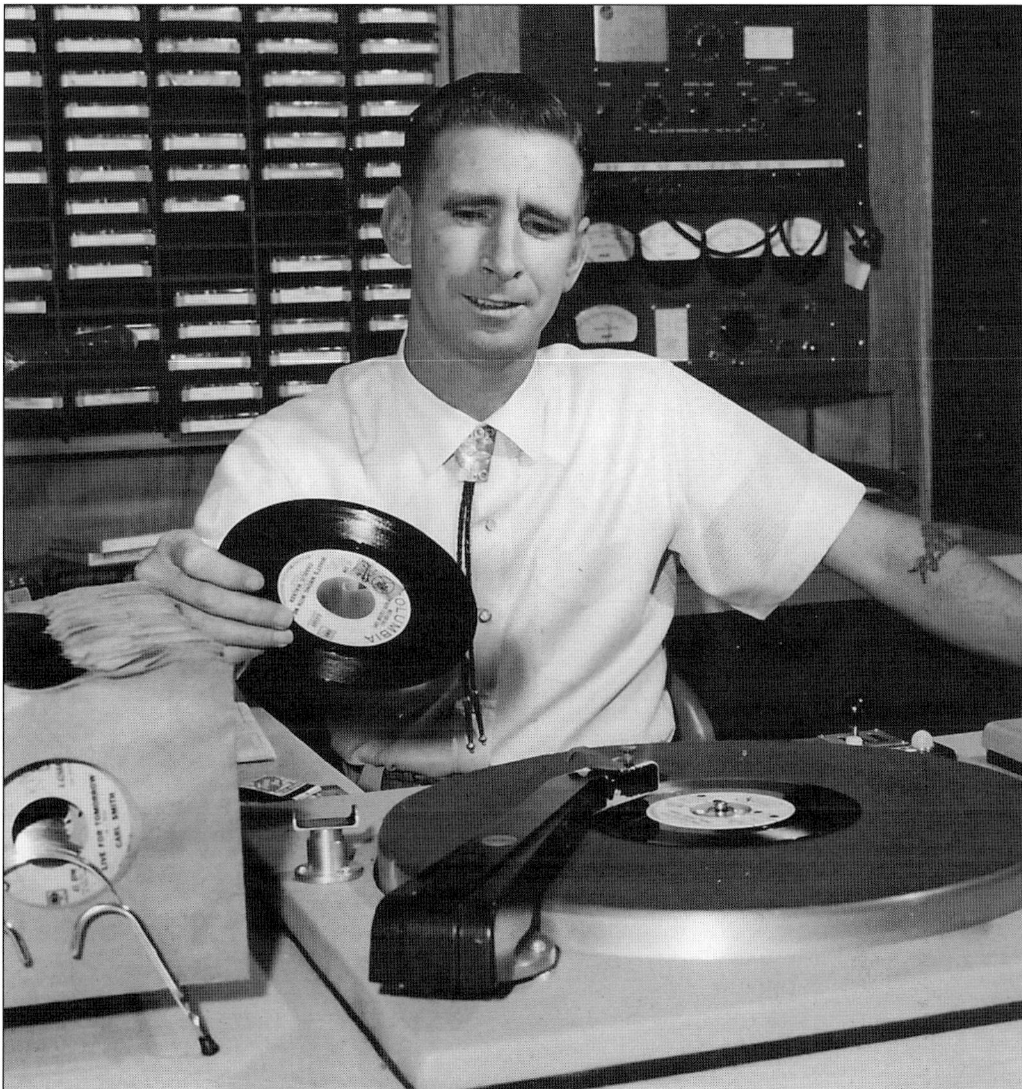

KMOP 1330 came on the air in 1957. It was owned by country radio personality Ray Odom, who had started country music stations KHEP and KHAT in Phoenix. The first KMOP studio was in the Tucson Inn on West Drachman Street, where Odom did an on-air shift in addition to managing the station. He sold the station to the Clayton family in 1959 and returned to Phoenix. An early KMOP personality was the talented "Bongo Bob" Wolfe, who came over from KTKT in 1959. Smokey Stover was hired by station manager Ken Ferguson in 1960. Here, he is at the turntables in the new KMOP studio on West Grant Road. His rack of records contains the "Live 45," the current popular hits. Stover's folksy approach to morning radio made him very popular. He and newscaster Eddie Jackson did a "farm report" each morning, assuming the characters "Lonnie and Gene" and telling jokes while reporting the news. Hourly news headlines were called the "Shotgun" edition, with each story punctuated by a (recorded) gunshot. Stover also appeared on weekends at a small country music nightclub, where his listeners could meet him and hear him sing some of his songs, many of which were recorded by George Jones and other country stars. After leaving Tucson, Stover worked at other radio stations. In 2000, he was elected to the Country Music Disc Jockey Hall of Fame. (RJ.)

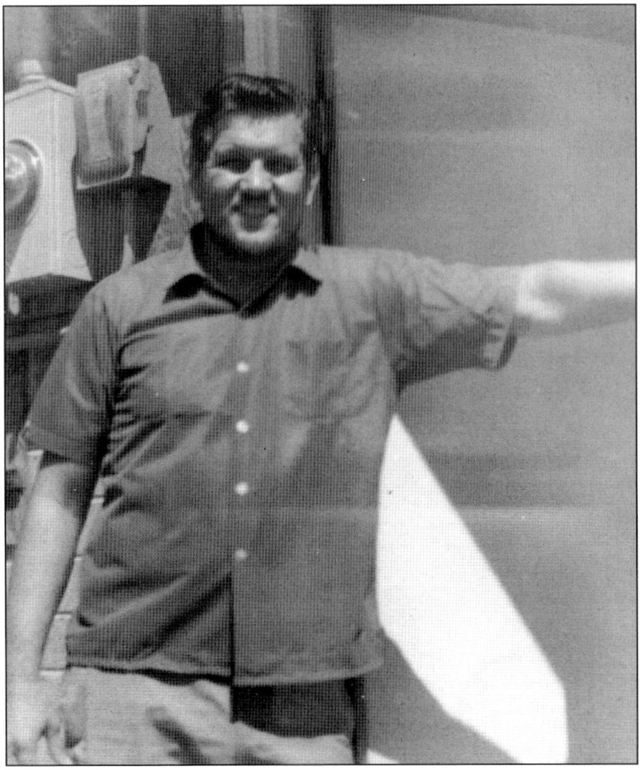

Eddie Jackson was one of those people who loved being on the radio, but remained working full-time in another occupation. Jackson did his weekend program in addition to being Smokey Stover's newscaster before heading to work at the San Manuel mine. When he was available, KMOP brought Jackson in to work at the studio while the regular personality did his show from a remote location or at a station-sponsored event. Eddie's theme song for his Sunday program was "Bud's Bounce," a lively instrumental featuring the pedal steel guitar of Bud Isaacs, the legendary sideman who appeared on over 100 records. (RJ.)

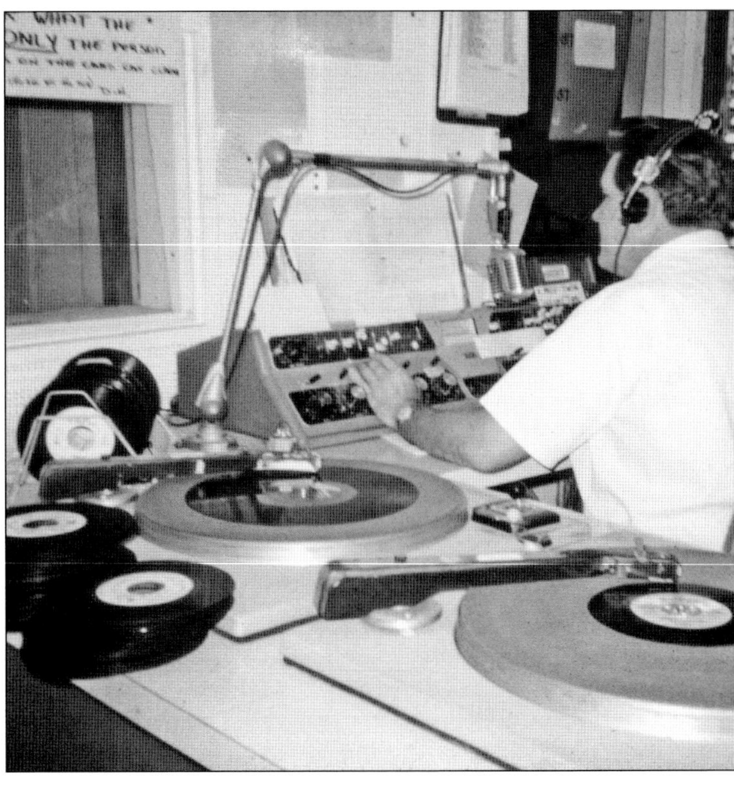

Eddie Jackson is seen here on the air in the KMOP control room. In addition to his KMOP radio job and subsequent work on KHYT, Jackson's mining career broadened when he was elected an official of the union. He participated in many political campaigns and events and once ran for the Arizona legislature. Jackson had deep roots in Arizona, coming from Superior, where his family were miners. His wife, Shirley, and their three children remain in the state. His son Jeffrey was a writer-editor for the *Daily Star* for many years. (Shirley Jackson.)

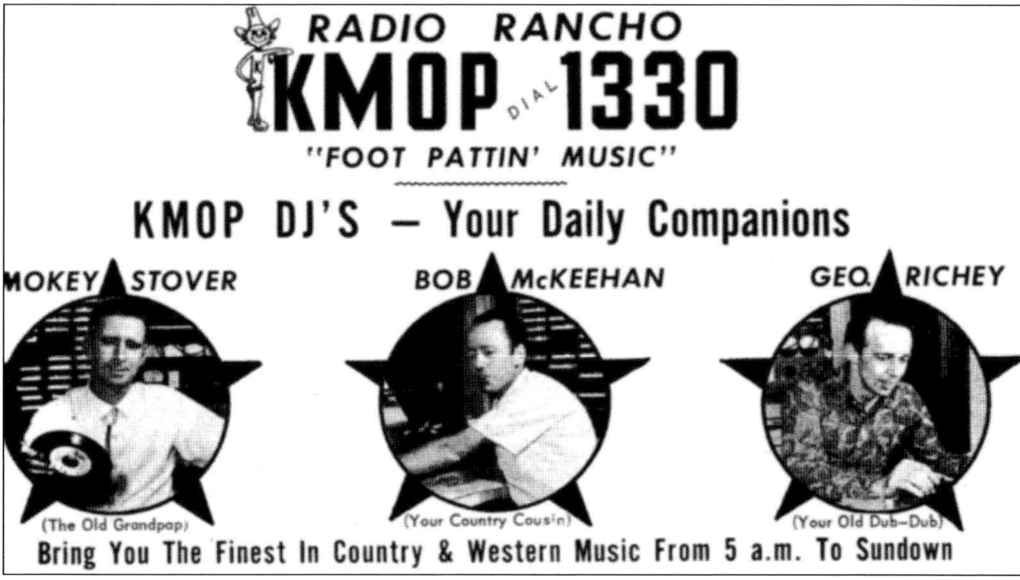

George Richey came to Tucson after his stint in the US Army at Fort Huachuca. When he landed at KMOP, he could utilize his heritage in country music. Richey had played gospel piano on the Springfield, Missouri, country show *Ozark Jubilee*, which starred Red Foley. In Tucson, he started a recording studio and produced songs for local artists. In 1963, Richey formed a band, with himself on piano, and appeared at a new Tucson nightclub, often participating in KMOP-sponsored events. One of the guest stars to appear was Bobby Bare, who at the time had the big hits "Detroit City" and "500 Miles." As KMOP's program director and afternoon personality, Richey selected the songs to be played. When he and his family left Tucson, it was for a job in Seattle, then to Los Angeles on KGBS and producing records for Bonnie Guitar and other artists. (RJ.)

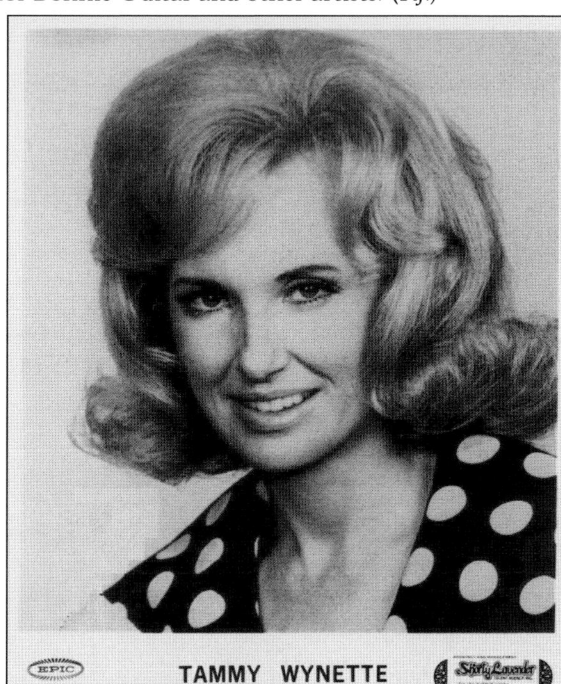

In 1967, Richey moved to Nashville and went on to write and produce songs for many top country singers. It was there that he met country music superstar Tammy Wynette and became her fifth husband. He directed her career from then on. Tammy's hits included the anthem "Stand By Your Man," "D-I-V-O-R-C-E," and "Apartment #9." As a sideline, Richey was music director for the television show *Hee Haw*. He and Wynette frequently returned to Tucson so she could have medical treatment. On one occasion, she did a benefit concert at the Maverick. (RJ.)

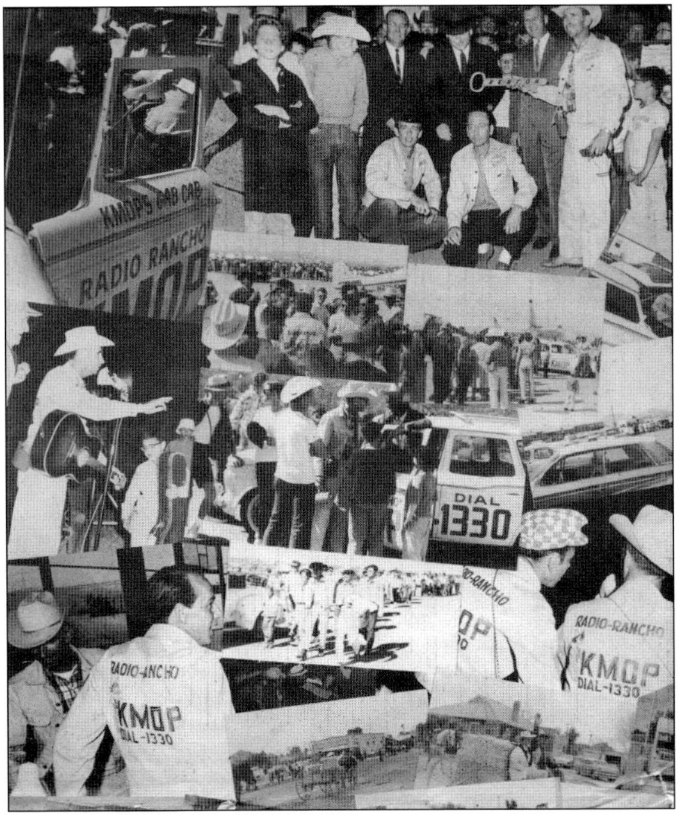

This collage features some of the events KMOP sponsored in the community. Bob McKeehan, George Richey, and Smokey Stover, complete with cowboy hats and guitars, were everywhere. A photograph at lower center shows the KMOP "Walk for Your Health" promotion, in which fans were challenged to walk to Benson along the old Benson highway. Eddie Jackson was at the station playing the records all day while the staff, stars, and fans walked. Most didn't finish. (RJ.)

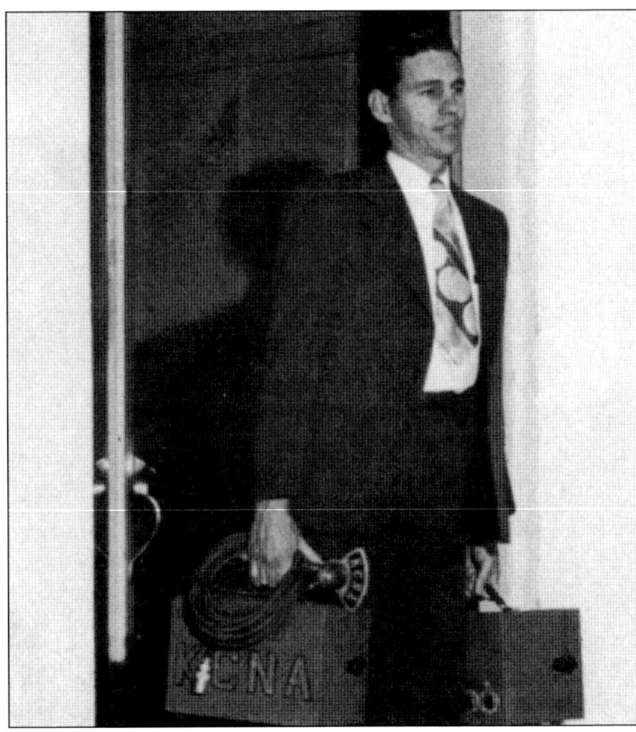

Ken Ferguson was an FCC-licensed engineer. When he was hired at KCNA, he was the first engineer who also did announcing. Ferguson moved into station management and eventually owned stations in Texas and Yuma before returning to Tucson in 1959 as KMOP's manager under the ownership of the Clayton family. The KMOP studios were moved from the Tucson Inn to a new location on West Grant Road, which Ferguson built. He left Tucson to own, manage, and engineer the only radio station outside Fort Huachuca, KHFH in Sierra Vista. Ferguson, originally from Casa Grande, was also a licensed pilot. (CJB.)

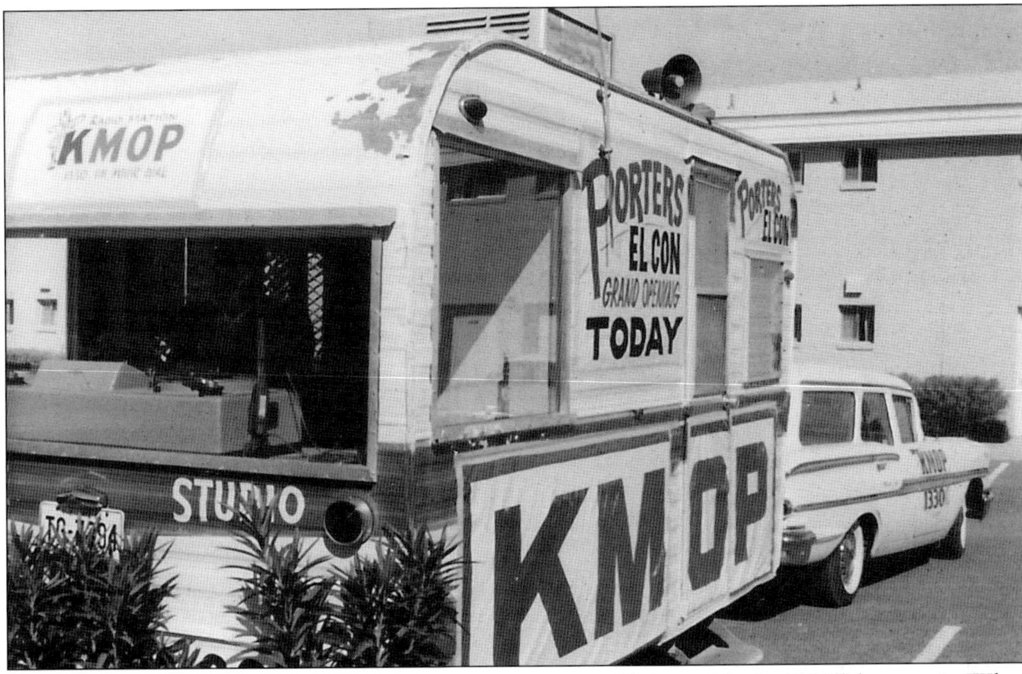

If this trailer looks familiar, it is because it is seen on page 60, but with the KAIR logo on it. When KAIR retrenched in the early 1960s, it sold the trailer to KMOP's Ken Ferguson, and it was used extensively to take the station to the audience for the openings of new stores and other events. KMOP personalities and musicians appeared and performed at the events. Bill Mortimer, who was hired to replace George Richey, was the last DJ to broadcast from the KMOP trailer. (RJ.)

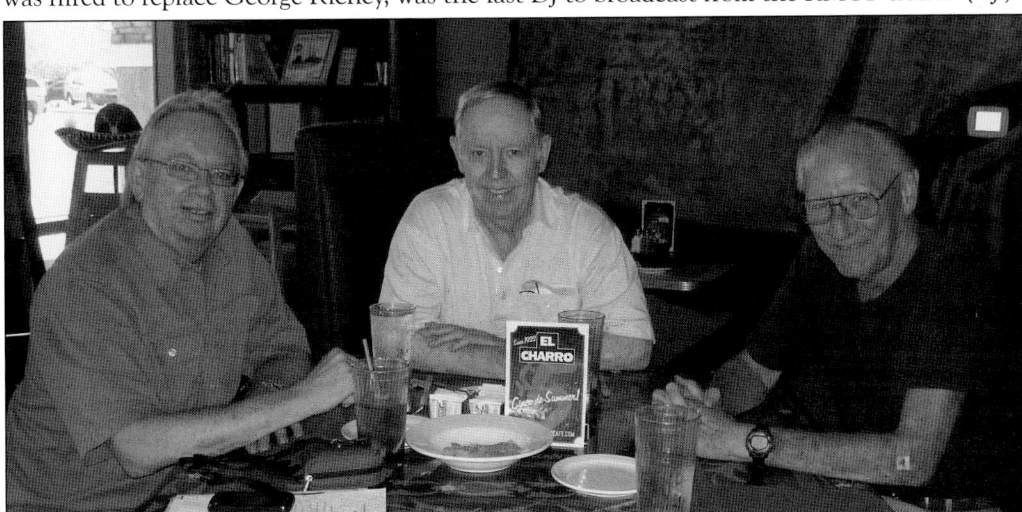

A trio of Tucson radio veterans reminisce about their careers. From left to right are Bill Mortimer, Jim Slone, and newscaster Don Wallace. Mortimer is a Tucson-area native who grew up on the Canoa Ranch at Green Valley. He still lives there, where his dad and grandparents worked through the 1940s and 1950s. Mortimer says, "Years later, when I was on Slone's KCUB, George Richey and Tammy Wynette stopped by the studios on Roger Road, and they both sat with me on the air for over an hour. It was fun, and I asked her about her years with George Jones. She was quite open to any topic." (Jim Slone.)

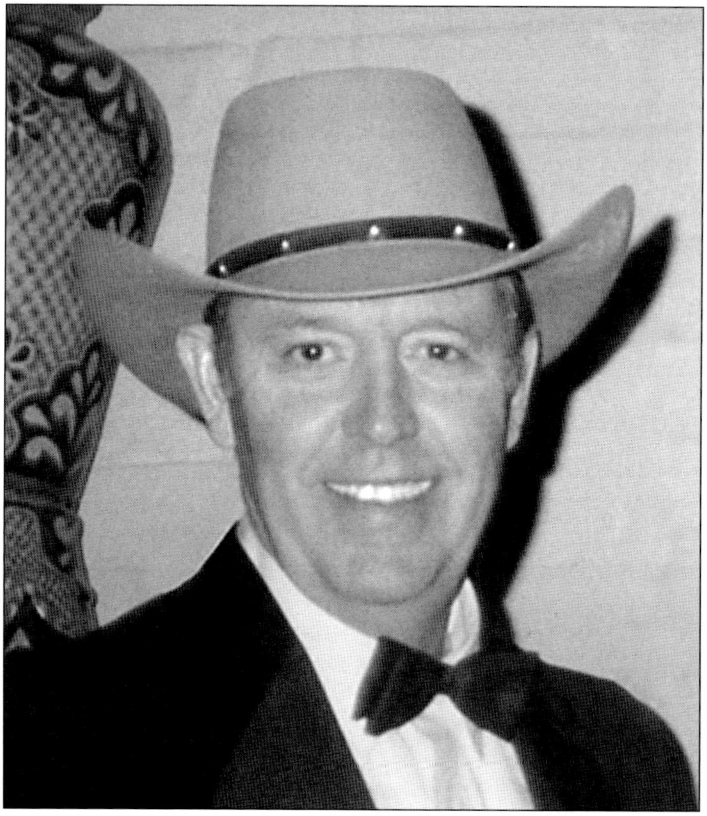

The Jim Slone era in Tucson began when he was hired by Fred Vance, the owner of the soon-to-be KHOS 940. Slone was in Albuquerque, but he came to Tucson to program the new 24-hour country music station. He came to town "in my 1956 Chevy ready to go as the morning man when I was 26 and have been here ever since. . . . All in all, I couldn't have asked for better. Was I apprehensive? No. Excited? YES." KHOS signed on the air on August 10, 1963, at 7:15 p.m., just as KMOP signed off for the day. The first song Slone played was Porter Wagoner's 1954 hit, "Company's Comin'." (Jim Slone.)

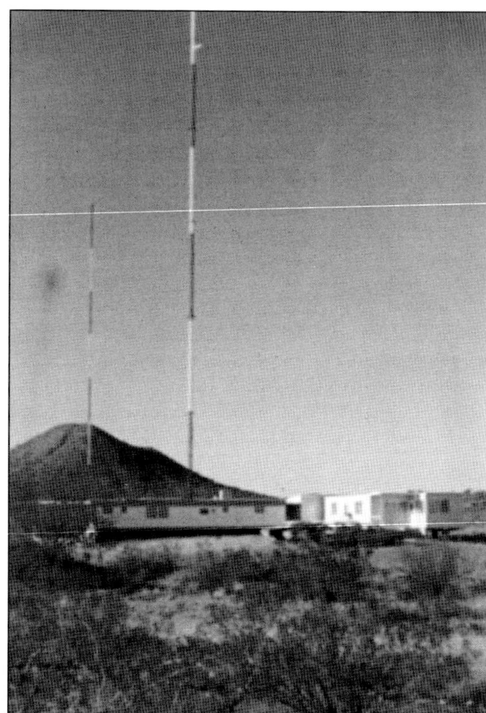

This is where KHOS was located when it signed on the air in 1963, on San Juan Trail southwest of "A" Mountain. The studio was in the mobile home on the right, with the transmitter in the small building in the center. Chief engineer Walter Stiles and his wife lived in the mobile home on the left. Station owner Fred Vance, who was also the general manager of KVOA-TV, owned two-thirds of KHOS. Vance became sole owner in 1968. (Jim Slone.)

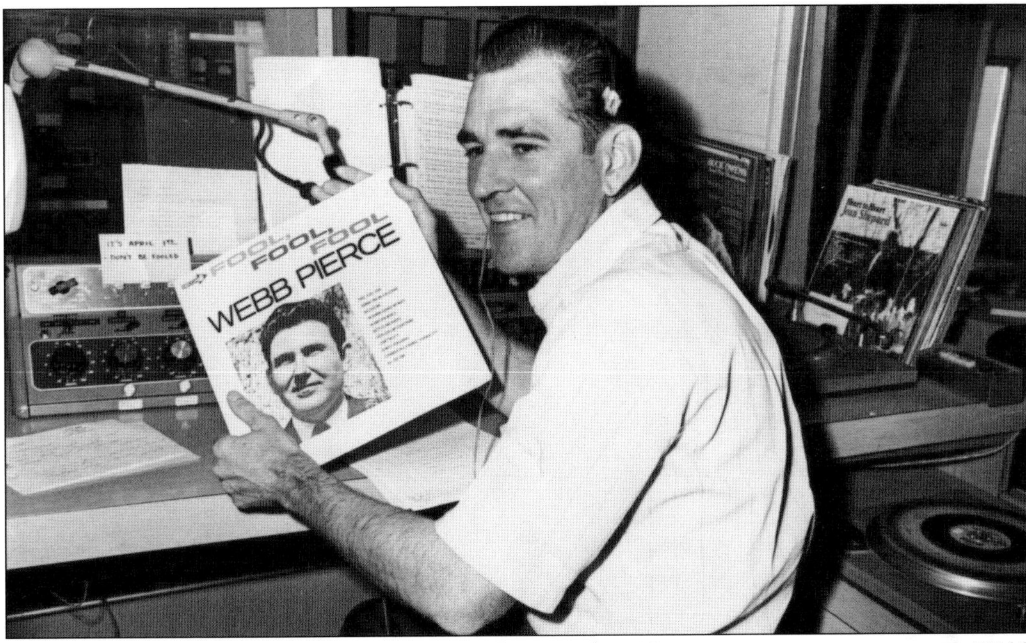

Sometimes, stations did cross-promotions with one other. Here, country music superstar Webb Pierce's album is held by Jerry Stowe, who, on that day in 1968, was sitting in at KHOS instead of doing his show on crosstown Top 40 KTKT. Yes, it was April 1, April Fool's Day. Jim Slone crossed over and did his show on KTKT in place of Jimmy Rabbit, who came over to KHOS. "I was surprised they would agree," Slone says, "because we were putting heat on them in the ratings and they always ignored us." (Jerry Stowe.)

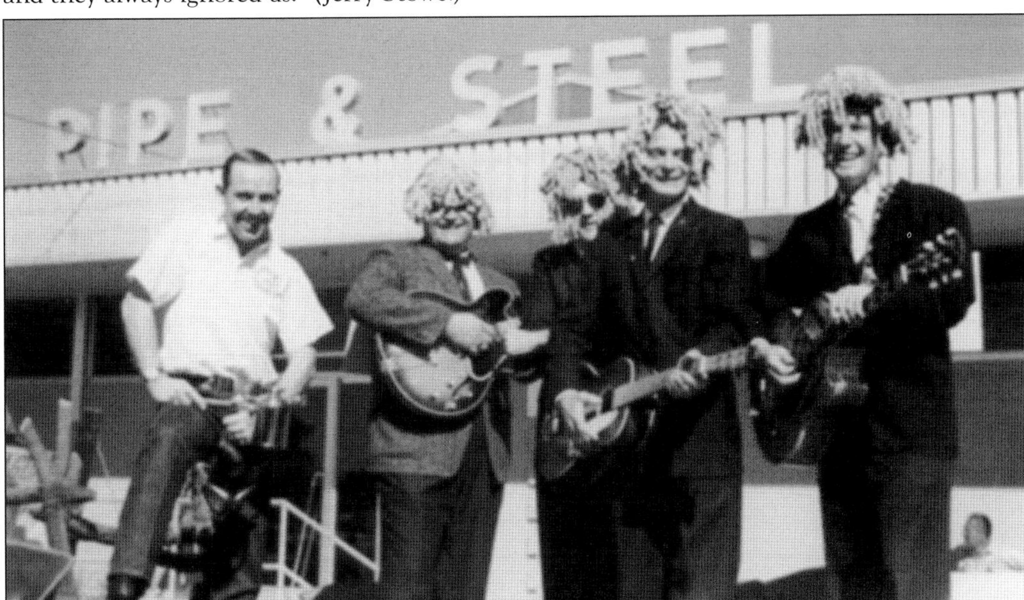

It was a given that radio people would appear in the Tucson Rodeo parade. This group of KHOS personalities, dressed as The Beatles, rode in the parade in 1965. They are, from left to right, Jim Slone, Rusty Wright, Bob Wolf, Ty Conner, and Chuck Hall. The guys are ready to entertain the folks along the parade route. (Jim Slone.)

KHOS 94

PREVIEWED OCTOBER 25, 1969

NO.	TITLE	ARTIST	LABEL
1	The Ways To Love A Man	Tammy Wynette	Epic
2	Since I Met You Baby	Sonny James	Capitol
3	To See My Angel Cry	Conway Twitty	Decca
4	To Make A Man (Feel Like A Man)	Loretta Lynn	Decca
5	These Lonely Hands of Mine	Mel Tillis	Kapp
6	I'd Rather Be Gone	Hank Williams, Jr.	MGM
7	Homecoming	Tom T. Hall	Mercury
8	George (And The North Woods)	Dave Dudley	Mercury
9	Are You From Dixie	Jerry Reed	RCA
10	Invitation To Your Party	Jerry Lee Lewis	Sun
11	That's A No No	Lynn Anderson	Chart
12	I Will Always	Don Gibson	RCA
13	Muddy Mississippi Line	Bobby Goldsboro	UA
14	Okie From Muskogee	Merle Haggard	Capitol
15	Seven Lonely Days	Jean Shepard	Capitol
16	She Even Woke Me Up To Say Goodbye	Jerry Lee Lewis	Smash
17	Such A Fool	Roy Drusky	Mercury
18	Life's Little Ups And Downs	Charlie Rich	Epic
19	Don't It Make You Want To Go Home	Joe South	Capitol
20	I'll Still Be Missing You	Warner Mack	Decca
21	We All Go Crazy	Jack Reno	Dot
22	Kissed By The Rain, Warmed By The Sun	Glenn Barber	Hickory
23	Then The Baby Came	Henson Cargill	Monument
24	Little Reasons	Charlie Louvin	Capitol
25	Another Day, Another Mile, Another Highway	Clay Hart	Metromedia
26	Back In The Arms Of Love	Jack Greene	Decca
27	The House Of Blue Lights	Earl Richards	UA
28	September Song	Roy Clark	Dot
29	Better Homes And Gardens	Billy Walker	Monument
30	New Orleans	Anthony A. Jones	Chart
31	Ship In The Bottle	Stonewall Jackson	Columbia
32	Groovy Grubworm	Harlow Wilcox	Plantation
33	Where Have All The Average People Gone	Roger Miller	Smash
34	My Blue Ridge Mountain Boy	Dolly Parton	RCA
35	My Iron Skillet	Wanda Jackson	Capitol
36	There Wouldn't Be A Lonely Heart In Town	Del Reeves	UA
37	We Had All The Good Things Going	Jan Howard	Decca
38	While I'm Thinkin' It	Billy Mize	Imperial
39	Try A Little Kindness	Glen Campbell	Capitol
40	Thank You For Loving Me	Brenda Byers	MTA

KHOS TWIN PICKS

	TITLE	ARTIST	LABEL
●	Afraid Of Losing You Again	Charley Pride	RCA
●	Big In Vegas	Buck Owens	Capitol

After Slone became manager, KHOS was, from then on, always pronounced "K-HOSS." The station dropped the description of "home folks radio," but used a bucking horse (hoss) as its symbol. This is a KHOS 1969 song catalog, listing the country songs played. The list includes many top country artists of the day, with Tammy Wynette in the top position. Also appearing are Sonny James, Conway Twitty, Loretta Lynn, Don Gibson, Charlie Rich, Charlie Louvin, Jack Greene, Roy Clark, Dolly Parton, Wanda Jackson, Buck Owens, and Charlie Pride. (CJB.)

When Tuscan country music fans wanted to see their favorite stars in person, they could find them appearing at the Tucson Gardens on West Lester Street. Country music stations sponsored these events. One of the first concerts promoted by KHOS was Hank Thompson and his great band in 1964. Johnny Cash, Buck Owens, and Willie Nelson also performed there. George Jones was scheduled for two concerts promoted by KHOS, and Jim Slone remembers Jones did one of his no-shows at the first one. Refunds had to be given. After the second no-show, "we put a judgment on him that if he came into the state of Arizona we would garnish his proceeds." (CJB.)

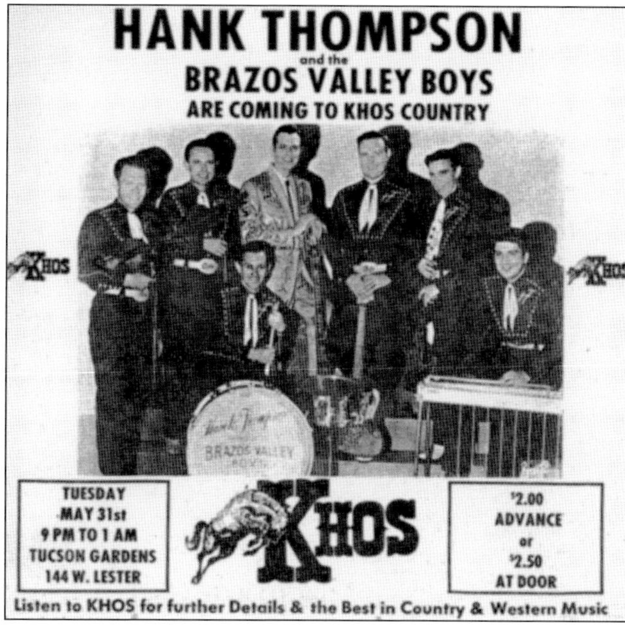

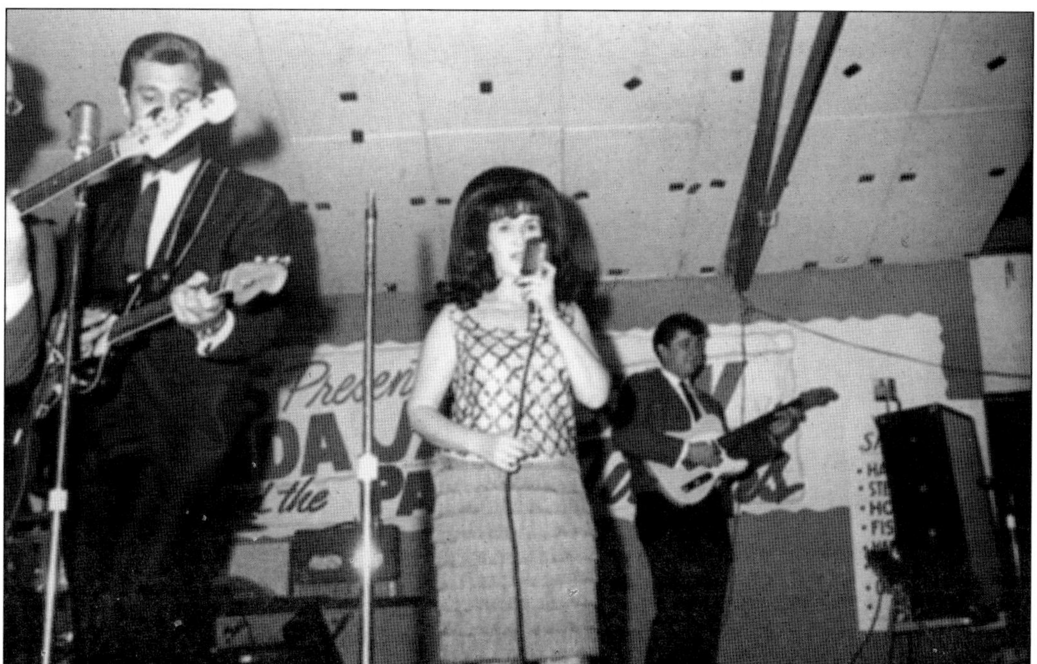

Here, Wanda Jackson appears at Tucson Gardens, sponsored by KHOS. (Jim Slone.)

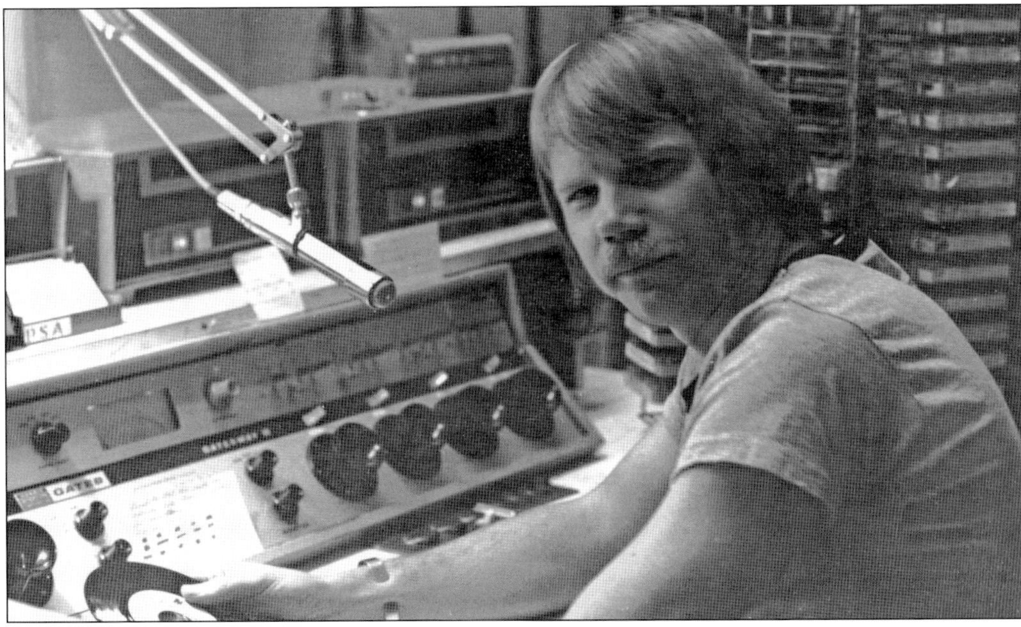

In 1971, KHOS was doing well, which attracted the interest of KCUB's owner Robert Chandler, whose station was near the bottom of the Tucson ratings. He approached Jim Slone and offered him the job of managing KCUB. Slone said he would do so if he could earn ownership. The station switched to country music, and the promotions featured the KCUB Bunch of outstanding radio personalities, like Bob Lee, seen here in 1972 at the control board. Within a few years, KCUB earned top ratings in town. (Bob Lee.)

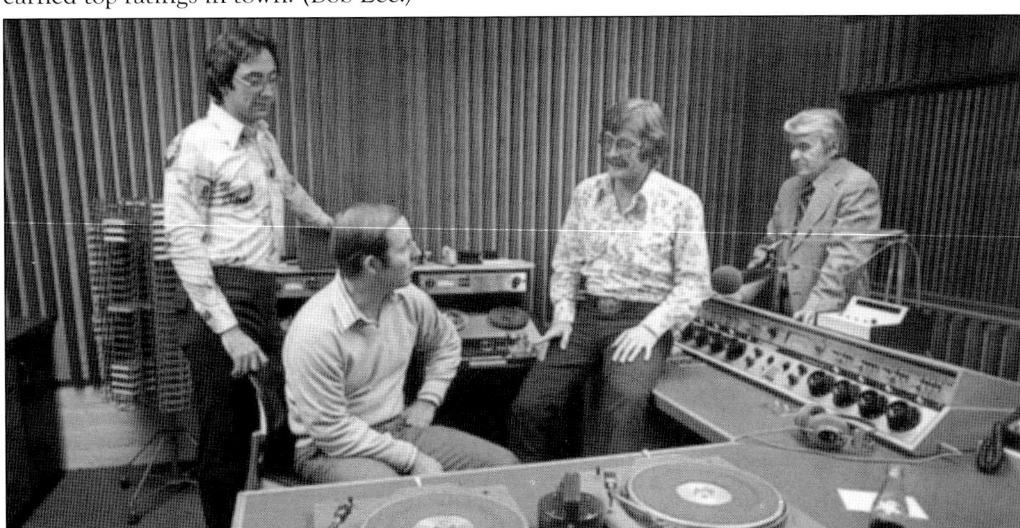

Many of the employees came with Slone to KCUB from KHOS. He hired former KTKT general manager Phil Richardson to serve as sales manager. Richardson also did the news reports on the hour. Slone allowed five minutes for the news, "so we could get back to the music in a hurry," but those newscasts leaned heavily on local events. Seen here in 1976 in the KCUB production room are, from left to right, program director Bob English, Jim Slone, morning personality "Sunny Jim" Arnold, and Phil Richardson. Jim Slone increased his share in KCUB until he owned it all. (*Arizona Daily Star* and Jim Slone.)

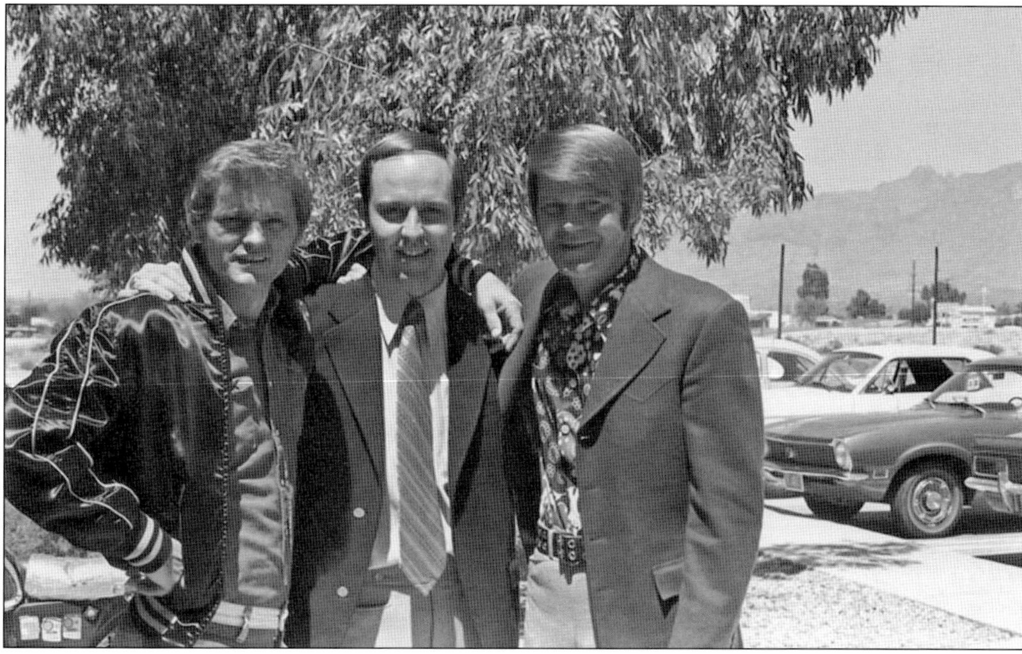

KCUB also hosted big country stars for appearances and station interviews. While at KHOS, in 1964, Jim Slone had brought the great guitar player and singer Glen Campbell to the Tucson Gardens. In 1972, Slone's first KCUB concert promotion brought country star Jerry Reed (left) and Campbell (right) to Tucson. Slone is the center of "three good ol' boys." (Jim Slone.)

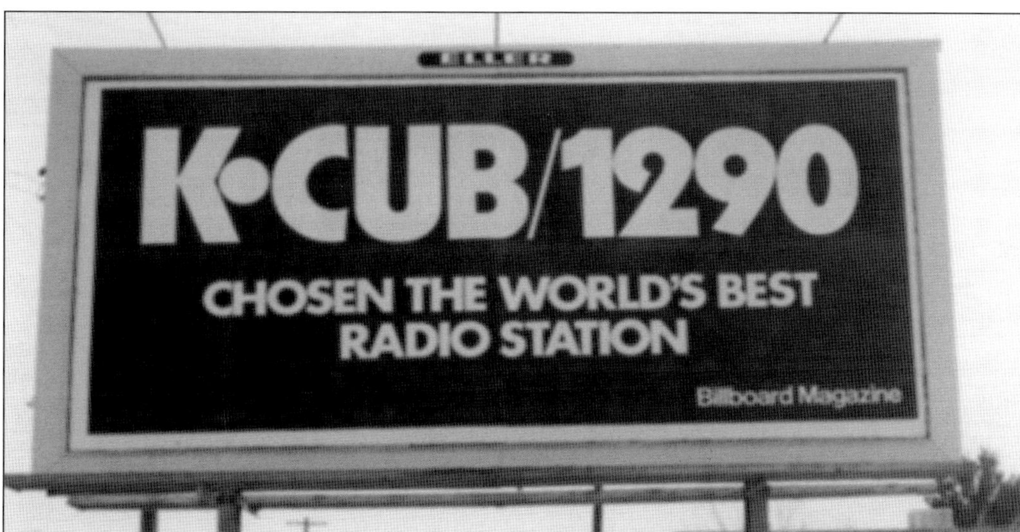

In December 1976, *Billboard* magazine recognized KCUB as the best radio station in the world! Slone and Jim Arnold attended the Billboard New Orleans Radio Convention and, upon their return, had these billboards put up all over town. That year, Sunny Jim was a finalist for air personality of the year. Slone's programming philosophy was to avoid the trend of playing exclusively modern country music, as KHOS had done. KCUB took the more traditional route, which was right for Tucson. (Jim Slone.)

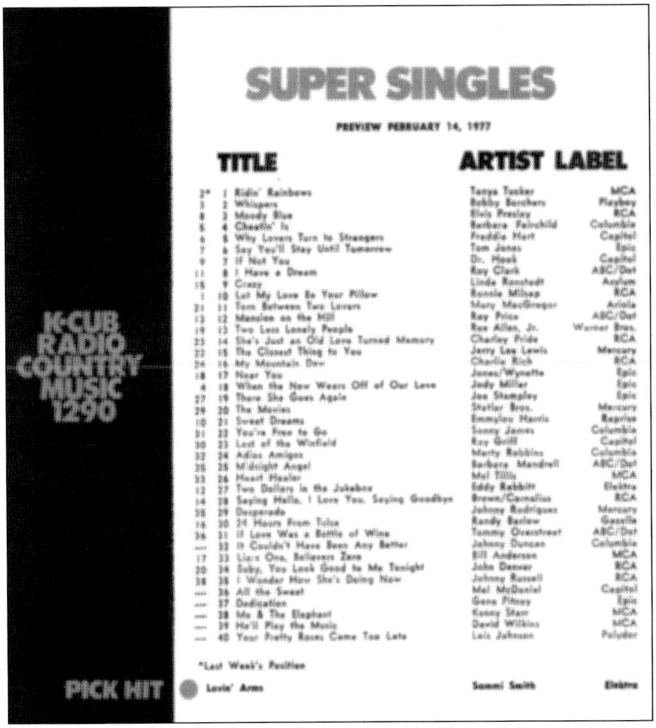

SUPER SINGLES

PREVIEW FEBRUARY 14, 1977

K-CUB RADIO COUNTRY MUSIC 1290

*		TITLE	ARTIST	LABEL
2	1	Ridin' Rainbows	Tanya Tucker	MCA
3	2	Whispers	Bobby Borchers	Playboy
8	3	Moody Blue	Elvis Presley	RCA
5	4	Cheatin' Is	Barbara Fairchild	Columbia
6	5	Why Lovers Turn to Strangers	Freddie Hart	Capitol
7	6	Say You'll Stay Until Tomorrow	Tom Jones	Epic
9	7	If Not You	Dr. Hook	Capitol
11	8	I Have a Dream	Roy Clark	ABC/Dot
15	9	Crazy	Linda Ronstadt	Asylum
1	10	Let My Love Be Your Pillow	Ronnie Milsap	RCA
21	11	Torn Between Two Lovers	Mary MacGregor	Ariola
13	12	Mansion on the Hill	Ray Price	ABC/Dot
19	13	Two Less Lonely People	Rex Allen, Jr.	Warner Bros.
23	14	She's Just an Old Love Turned Memory	Charley Pride	RCA
22	15	The Closest Thing to You	Jerry Lee Lewis	Mercury
24	16	My Mountain Dew	Charlie Rich	RCA
18	17	Near You	Jones/Wynette	Epic
4	18	When the New Wears Off of Our Love	Jody Miller	Epic
27	19	There She Goes Again	Joe Stampley	Epic
29	20	The Movies	Statler Bros.	Mercury
10	21	Sweet Dreams	Emmylou Harris	Reprise
31	22	You're Free to Go	Sonny James	Columbia
30	23	Last of the Winfield	Roy Griff	Capitol
32	24	Adios Amigos	Marty Robbins	Columbia
25	25	Midnight Angel	Barbara Mandrell	ABC/Dot
33	26	Heart Healer	Mel Tillis	MCA
12	27	Two Dollars in the Jukebox	Eddy Rabbitt	Elektra
14	28	Saying Hello, I Love You, Saying Goodbye	Brown/Cornelius	RCA
35	29	Desperado	Johnny Rodriguez	Mercury
16	30	24 Hours From Tulsa	Randy Barlow	Gazelle
36	31	If Love Was a Bottle of Wine	Tommy Overstreet	ABC/Dot
—	32	It Couldn't Have Been Any Better	Johnny Duncan	Columbia
17	33	Liars One, Believers Zero	Bill Anderson	MCA
20	34	Baby, You Look Good to Me Tonight	John Denver	RCA
38	35	I Wonder How She's Doing Now	Johnny Russell	RCA
—	36	All the Sweet	Mel McDaniel	Capitol
—	37	Dedication	Gene Pitney	Epic
—	38	Me & The Elephant	Kenny Starr	MCA
—	39	He'll Play the Music	David Wilkins	MCA
—	40	Your Pretty Roses Came Too Late	Lois Johnson	Polydor

*Last Week's Position

PICK HIT ● Lovin' Arms — Sammi Smith — Elektra

In 1977, KCUB won "Country Music Station of the Year" in *Billboard* magazine. This 1977 KCUB music chart shows some of the country music artists being played. The No. 1 record was by Tayna Tucker, with others by Freddie Hart, Roy Clark, Ronnie Millsap, a duet by George Jones and Tammy Wynette, the Statler Brothers, and the legendary Roy Acuff. It also included Elvis Presley, Linda Ronstadt, and Jerry Lee Lewis. Slone tells of being on a flight in which Elvis was also to be on board. He arranged with the hostess to have Elvis sit across the aisle, so he could meet him. In 1978, the KCUB Bunch of personalities still included Bill Mortimer and Jim Arnold, but also had Jim West, Doug Brannon, and Dave Slavin. (Jim Slone.)

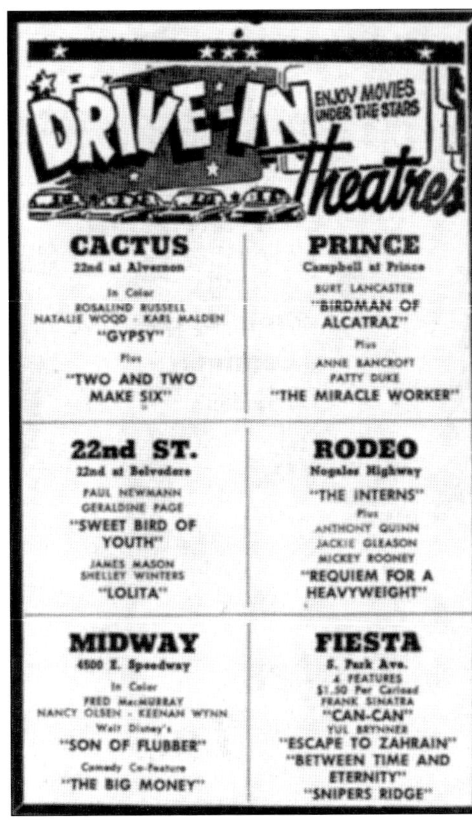

Radio was portable in the 1960s and 1970s, and it even went to the movies. In 1968, Tucson had six drive-in movie theaters, as shown in this ad. Each one showed a double feature each night, with folks watching the shows from their cars. Radio stations developed promotions with the theaters and gave away tickets. In 1968, KHOS had a big promotion going with the Midway on East Speedway. (Jim Slone.)

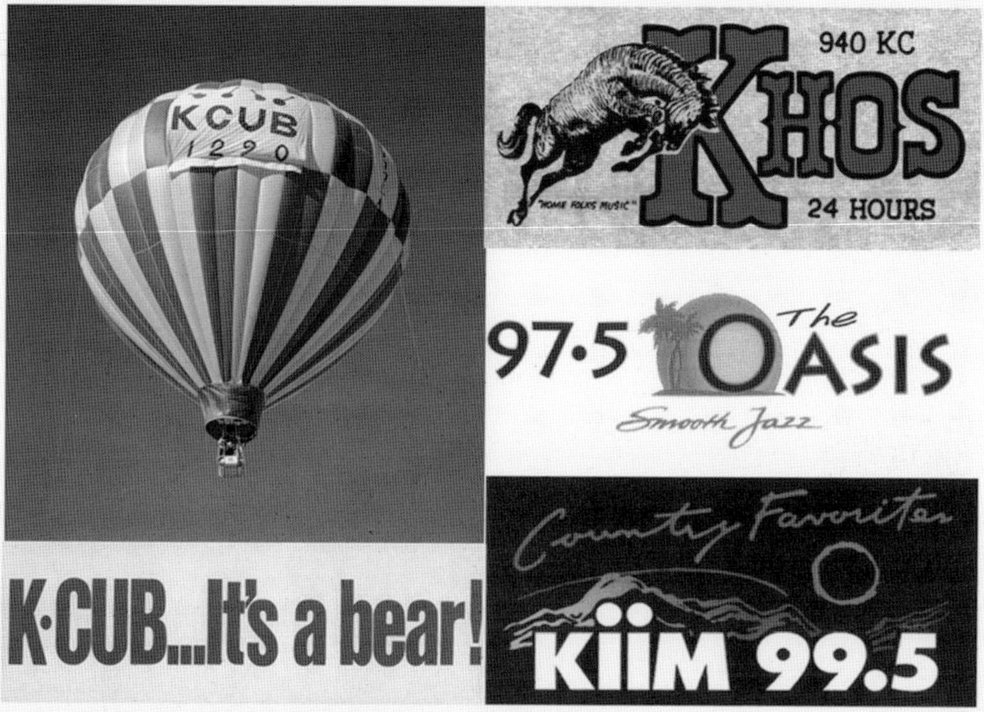

These are print logos and bumper stickers used by Slone's stations, starting with the KHOS bucking bronco in the upper right. The KCUB balloon floated all over town. The other logos were drawn by retired UA art professor Carl Heldt. "The Oasis" was a smooth-jazz format popular for a time on 97.5. KIIM 99.5's "Country Favorites" was the identifying phrase for country music on the radio in its move to FM. The "KCUB . . . It's a Bear!" bumper sticker resulted in 15,000 rolling billboards. Jim Slone told a 1983 *Tucson Citizen* interviewer, "Today, people are not afraid to say they listen to country music!" The Slone family was in business for the next 17 years, with sons Jamie and Fred in management and sales. Daughter Mary became cohost of the morning show on KIIM. (Jim Slone.)

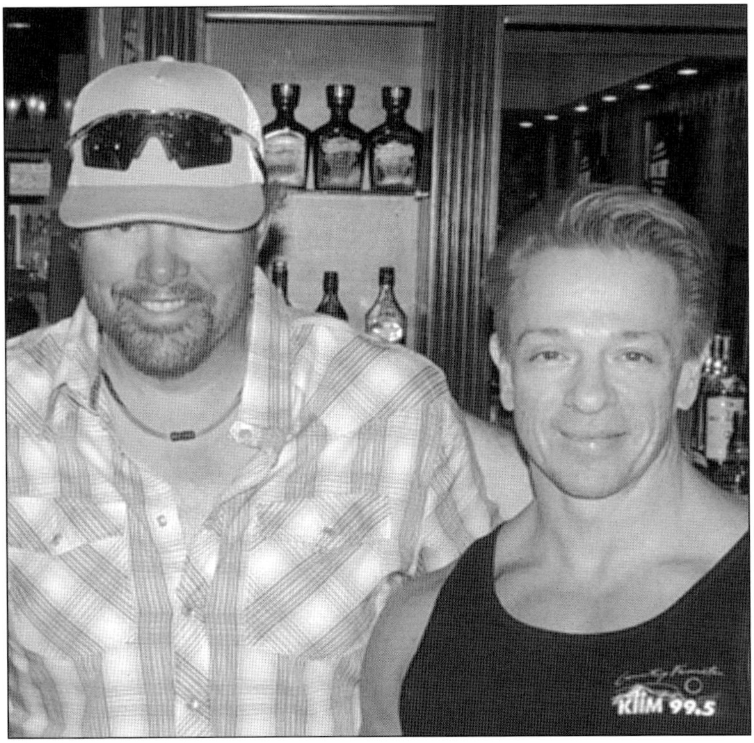

In 2000, Slone Broadcasting sold its FM stations KIIM, KHYT, KSZR, and the historic AM stations KTUC and KCUB. In 2001, Citadel Broadcasting hired a new face to do the morning show on KIIM-FM, christening the veteran broadcaster "Cactus Dave." He has held the job ever since, continuing the strong personality lineup on the country music station. In this photograph, Cactus Dave (right) is with country music superstar Toby Keith at the grand opening of I Love This Bar and Grill. (Cactus Dave.)

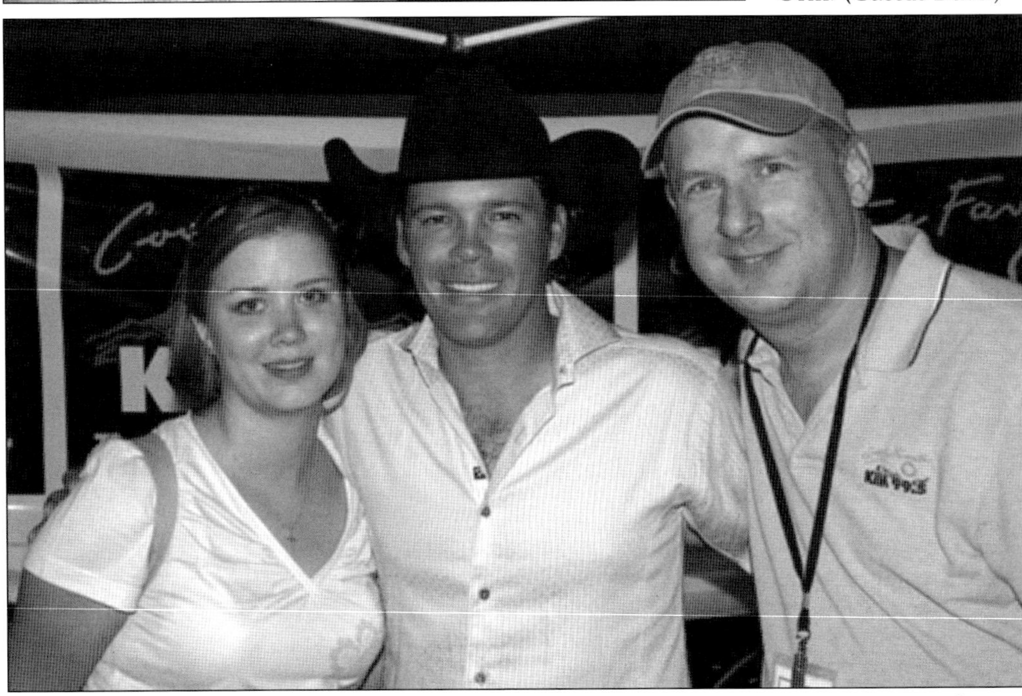

The tradition of country music stars coming to Tucson continued, as listeners expected to see their favorites in person and counted on the station and promoters to attract them. Here, longtime KIIM-FM program director Buzz Jackson (right) poses with superstar Clay Walker and Dena Walker in September 2007. (Buzz Jackson.)

Buzz Jackson is on the air in the KIIM-FM control room during a 2009 afternoon show. He is seen here on a promotion vehicle for Rascal Scooters. When asked if there were copies of music lists for the station, like the ones published in this book for other stations, Jackson said, "we haven't really published a playlist or country music chart for Tucson in the 12 years I've been here. . . . The Internet and website now carry that information!" Jackson has followed in the tradition of KIIM personalities, winning the Country Music Association Program Director of the Year award. (Buzz Jackson.)

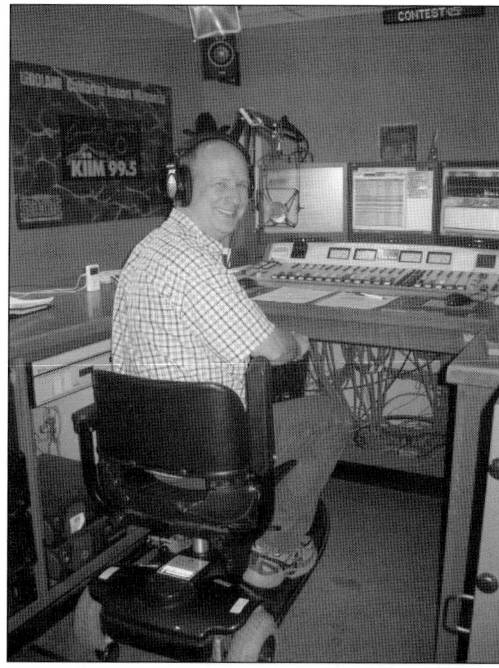

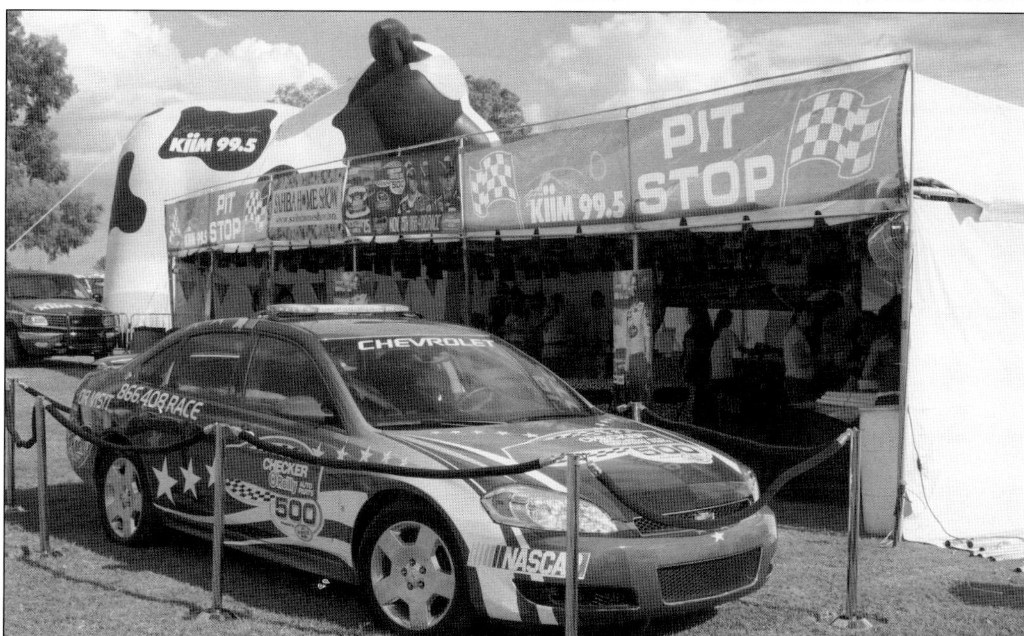

It is common for country music fans to also be NASCAR race fans. In this photograph, taken at the 2009 Tucson Country Music Festival sponsored by KIIM-FM, a Chevy is parked outside the station's booth. In the background, the inflatable KIIM cow towers over the scene. Jamie Slone, Jim's son, who stayed with the Citadel management for a while after his family sold it, is on the circuit as a professional racecar driver, as is his brother Fred. Citadel Broadcasting sold all of its radio stations, including the five in Tucson, to Cumulus Broadcasting, and the tradition carries on. (Buzz Jackson.)

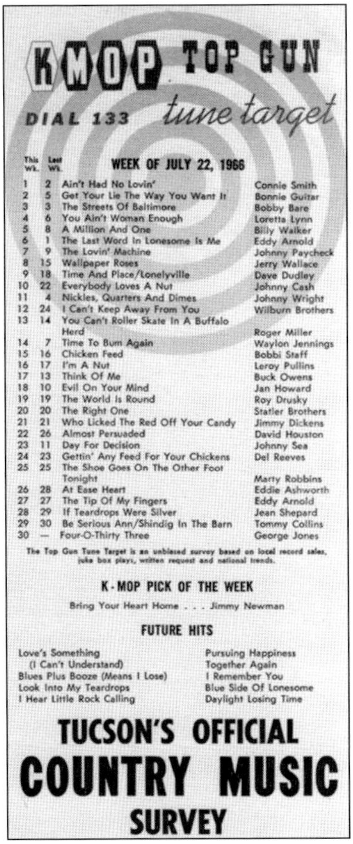

KMOP TOP GUN

DIAL 133 tune target

WEEK OF JULY 22, 1966

This Wk.	Last Wk.	Title	Artist
1	2	Ain't Had No Lovin'	Connie Smith
2	5	Get Your Lie The Way You Want It	Bonnie Guitar
3	3	The Streets Of Baltimore	Bobby Bare
4	6	You Ain't Woman Enough	Loretta Lynn
5	8	A Million And One	Billy Walker
6	1	The Last Word In Lonesome Is Me	Eddy Arnold
7	9	The Lovin' Machine	Johnny Paycheck
8	15	Wallpaper Roses	Jerry Wallace
9	18	Time And Place/Lonelyville	Dave Dudley
10	22	Everybody Loves A Nut	Johnny Cash
11	4	Nickles, Quarters And Dimes	Johnny Wright
12	24	I Can't Keep Away From You	Wilburn Brothers
13	14	You Can't Roller Skate In A Buffalo Herd	Roger Miller
14	7	Time To Bum Again	Waylon Jennings
15	16	Chicken Feed	Bobbi Staff
16	17	I'm A Nut	Leroy Pullins
17	13	Think Of Me	Buck Owens
18	10	Evil On Your Mind	Jan Howard
19	19	The World Is Round	Roy Drusky
20	20	The Right One	Statler Brothers
21	21	Who Licked The Red Off Your Candy	Jimmy Dickens
22	26	Almost Persuaded	David Houston
23	11	Day For Decision	Johnny Sea
24	23	Gettin' Any Feed For Your Chickens	Del Reeves
25	25	The Shoe Goes On The Other Foot Tonight	Marty Robbins
26	28	At Ease Heart	Eddie Ashworth
27	27	The Tip Of My Fingers	Eddy Arnold
28	29	If Teardrops Were Silver	Jean Shepard
29	30	Be Serious Ann/Shindig In The Barn	Tommy Collins
30	—	Four-O-Thirty Three	George Jones

The Top Gun Tune Target is an unbiased survey based on local record sales, juke box plays, written request and national trends.

K-MOP PICK OF THE WEEK

Bring Your Heart Home . . . Jimmy Newman

FUTURE HITS

Love's Something (I Can't Understand)
Blues Plus Booze (Means I Lose)
Look Into My Teardrops
I Hear Little Rock Calling
Pursuing Happiness
Together Again
I Remember You
Blue Side Of Lonesome
Daylight Losing Time

TUCSON'S OFFICIAL COUNTRY MUSIC SURVEY

When KHOS, with its 24-hour programming day, won the battle with the daytime-only KMOP, 1330 was sold to Californian Bob Scholz in 1965. The 1966 playlist chart at left shows that the station was still playing Loretta Lynn and Little Jimmy Dickens, who continued to perform at the Grand Ole Opry into their late 80s. Scholz applied for a full-time signal, but it was 15 frustrating years before that was granted. KHOS dominated country music radio in Tucson, leaving Scholz with no choice but to try something else, so he changed the call letters to KHYT and began playing Top 40 music. Below, Scholz (right) shows off his billboard to friends. The call letters "KHIT" were not available legally, but they could be referred to in publicity. To legally get that full-time signal, the station's city of license was changed to South Tucson. Barry Mishkind built the station's new transmitter near the Rincon Mountains and its new studio at Thirty-eighth and Eighth Streets. In 1980, 24-hour operation began. The KHYT Top 40 programming, led by Rich Brother Robbin, was very competitive. (Left, CJB; below, Barry Mishkind.)

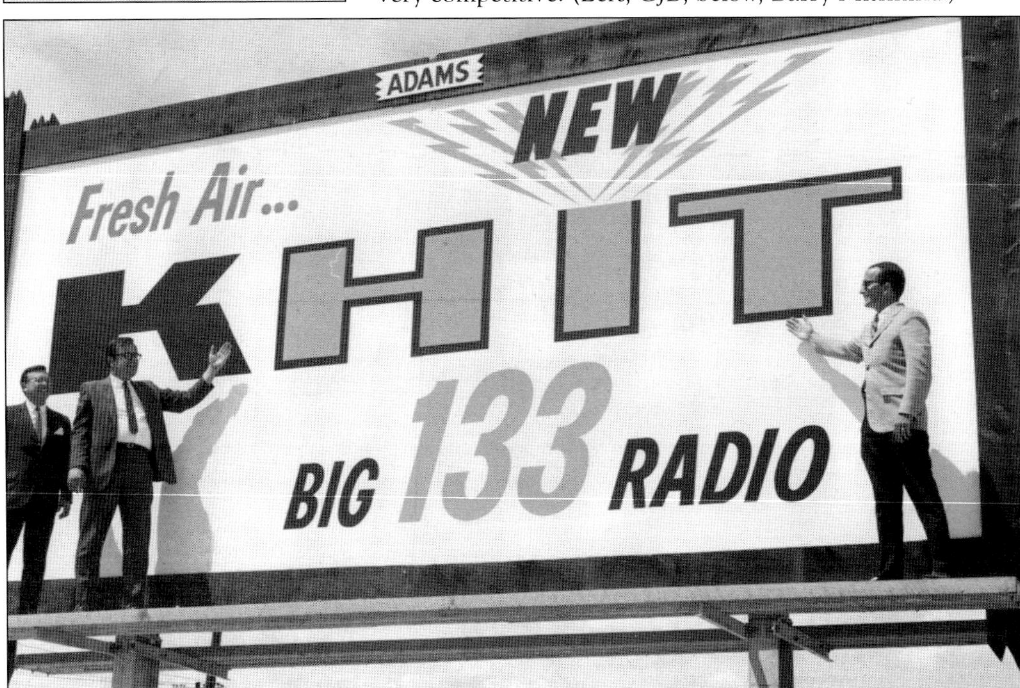

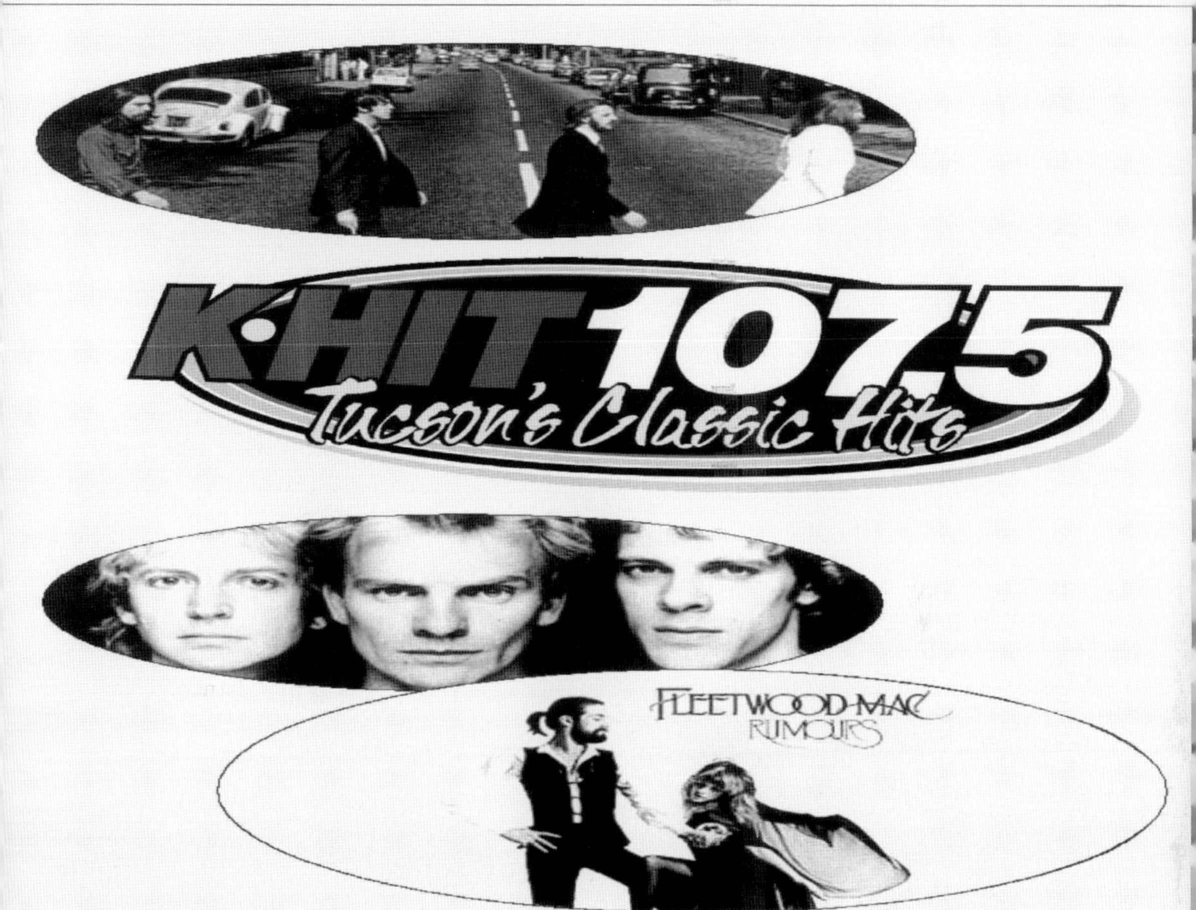

Although the call letters KHYT were changed to KMRR by new owners in 1987, no other station secured them until a new FM station was granted for Tucson at 107.5 in 1993, with its 100-kilowatt signal. Jim Slone saw an opportunity and purchased it from owner Tom Rockler in 1995. "I looked for available call letters and saw KHYT was available so selected them, but always called it K-HIT, 107.5 Classic Hits." The term "classic hits" was another name for rock "oldies," which has remained the format for the station throughout its history. It also carried UA football and basketball games on FM. As for country music, Jim Slone remains a fan of the music and its historic list of entertainers. (Lynn Spasoff, Cumulus Broadcasting.)

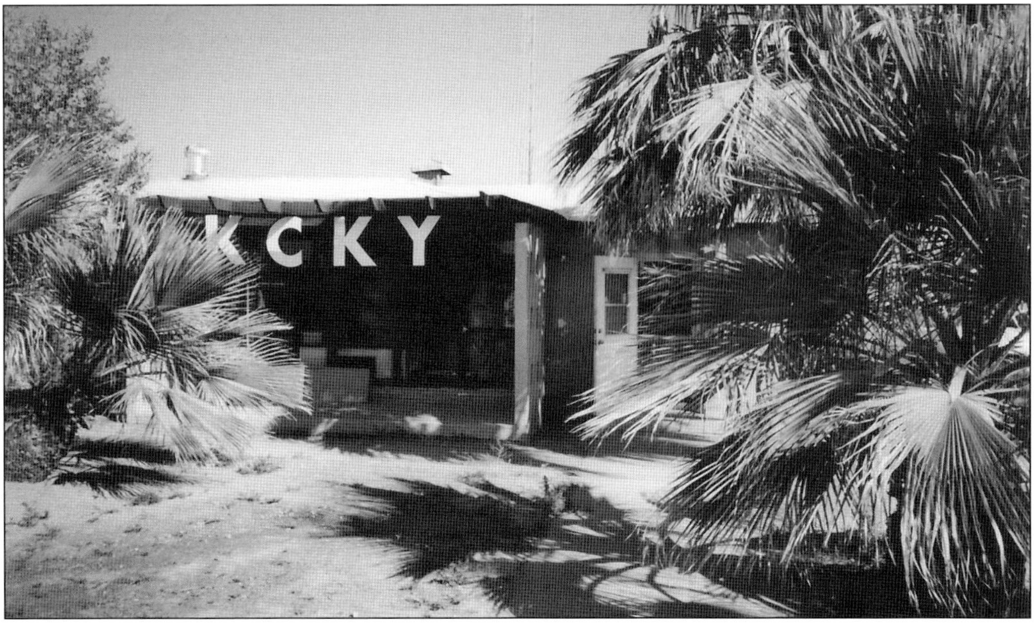

The fate of smaller community radio is reflected in what has transpired in Pinal County, just north of Tucson. With the growth of powerful AM and FM stations from Tucson and Phoenix, competing for local listeners became extremely difficult. At one time, there were five radio stations serving Pinal County with live local programming. One of the last to leave the air was KCKY 1150. It programmed country music and had local personalities, including future country music superstars Lee Hazelwood and Waylon Jennings. (RJ.)

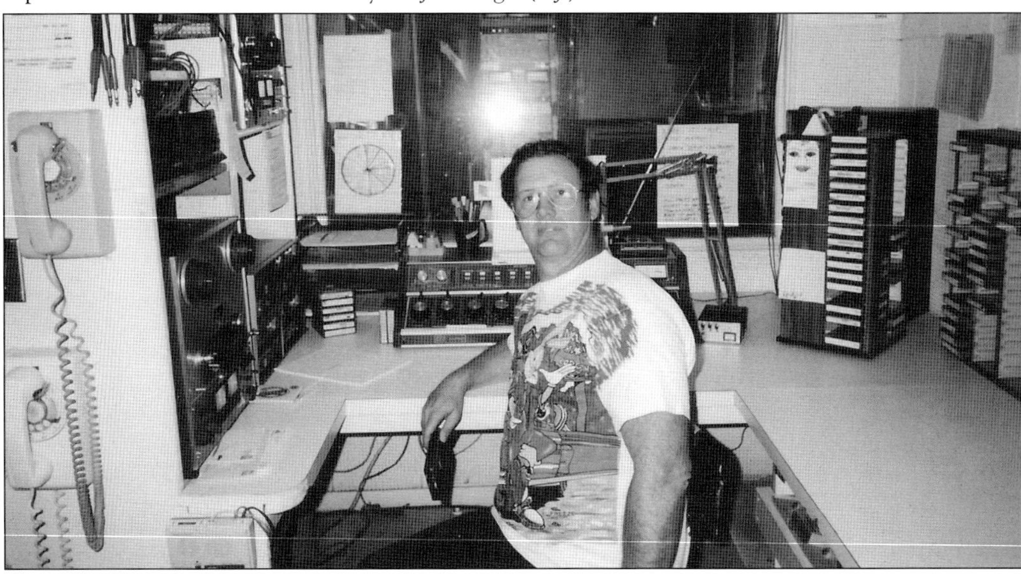

KCKY at one time was an affiliate of CBS Radio. KCKY's Coolidge studio was originally in the Studio Theater building downtown, but when that structure was torn down, it operated from this studio at the transmitter site north of town. The property had four houses where employees could live. KCKY became an original voice of sports talk and carried Chicago Cubs baseball. Bill Oostenburg (shown here) was the station's final employee in 1996 before the station became a Spanish-language religious station, ending radio originating in Pinal County. (RJ.)

Five

The Rest of the Story
News, Sports, Talk, Nostalgia, and the Hall of Fame

There is more to radio than music and commercials. The advertising pays the bills, keeps the stations on the air, and pays the employees, who in turn provide the programming listeners want to hear. People listen to radio because it is a companion and provides them with more than just a stack of records. When done right, it includes information and personality. News has been a part of radio's programming from the beginning; however, it took on an added importance during World War II. As television began to lure away large numbers of listeners, news reporting became radio's strongest asset. Tucson radio had received newscasts from national and Phoenix-based networks, but when "music and news" stations emerged, they were without network affiliation. Local news was soon delivered by local announcers, many of whom were also disc jockeys. Some stations sent newscasters and mobile units around town to acquire stories.

Talk programs, introduced in the 1980s, allowed the audience to participate in discussions of current issues. Public radio provided a national news and talk service. In larger cities, all-news stations provided continuous news coverage, but that was an expensive operation that Tucson stations could not afford. The weather forecasts, the time of day, and traffic reports all became important items. Sports events, reports, and fan discussions came along for that special audience. Nostalgia formats, in which a station played only songs that evoked long-ago memories, became popular. Radio stations were now able to develop a targeted audience and program to that niche. As the Internet and technology progressed, stations adapted to changing market conditions of when and where listeners were tuned in. Wrap all this around the personalities of those who were on the air and running the stations, and a group of individuals who contributed to the success of Tucson's radio industry emerges. Many of them are cited in this book and have been honored by being named to the Arizona Broadcasters Hall of Fame.

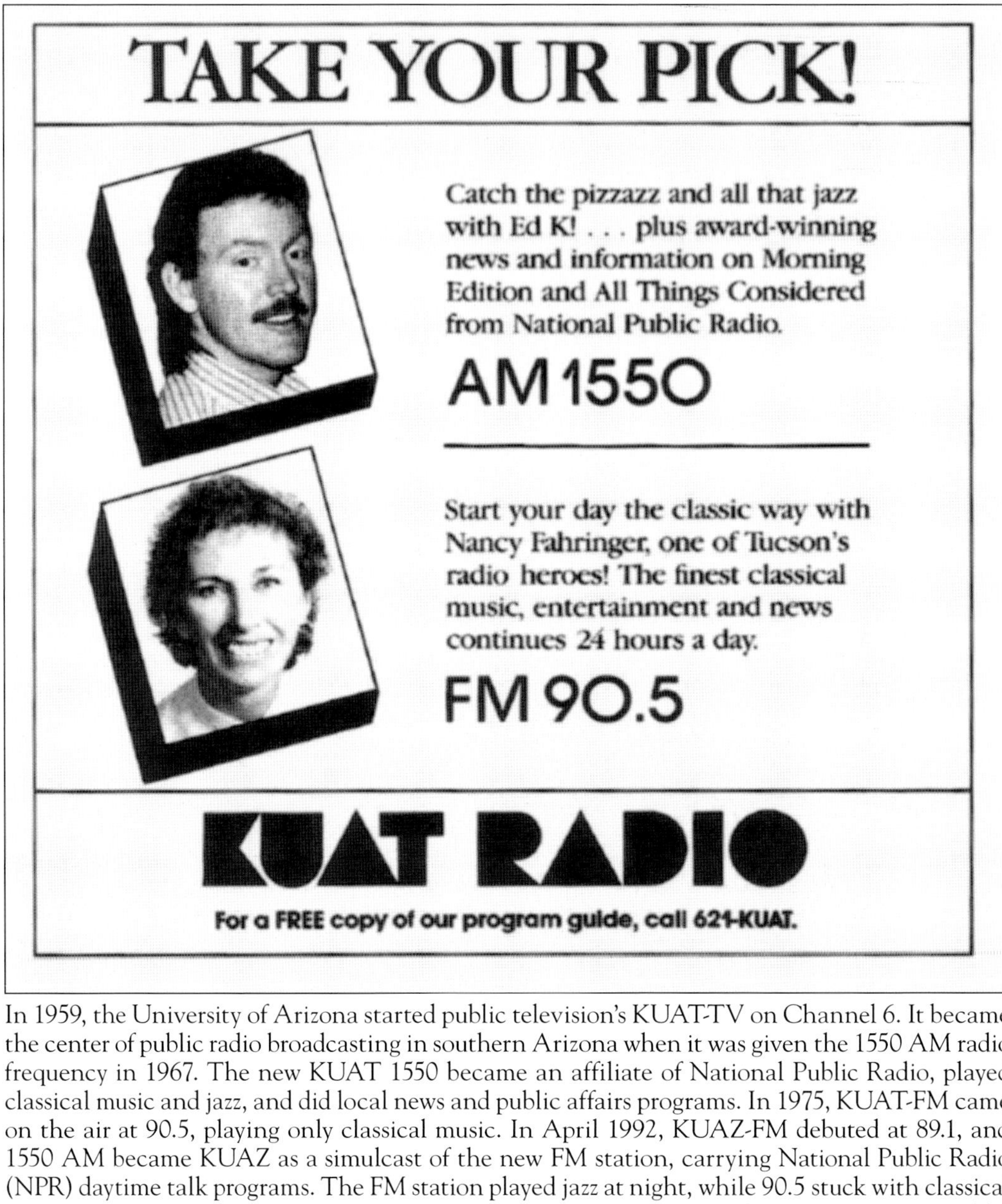

In 1959, the University of Arizona started public television's KUAT-TV on Channel 6. It became the center of public radio broadcasting in southern Arizona when it was given the 1550 AM radio frequency in 1967. The new KUAT 1550 became an affiliate of National Public Radio, played classical music and jazz, and did local news and public affairs programs. In 1975, KUAT-FM came on the air at 90.5, playing only classical music. In April 1992, KUAZ-FM debuted at 89.1, and 1550 AM became KUAZ as a simulcast of the new FM station, carrying National Public Radio (NPR) daytime talk programs. The FM station played jazz at night, while 90.5 stuck with classical music and some newscasts. This 1980s print advertisement features longtime Arizona Public Media radio program director Ed Kesterson and former classical morning host Nancy Fahringer. (John Kelley, Arizona Public Media.)

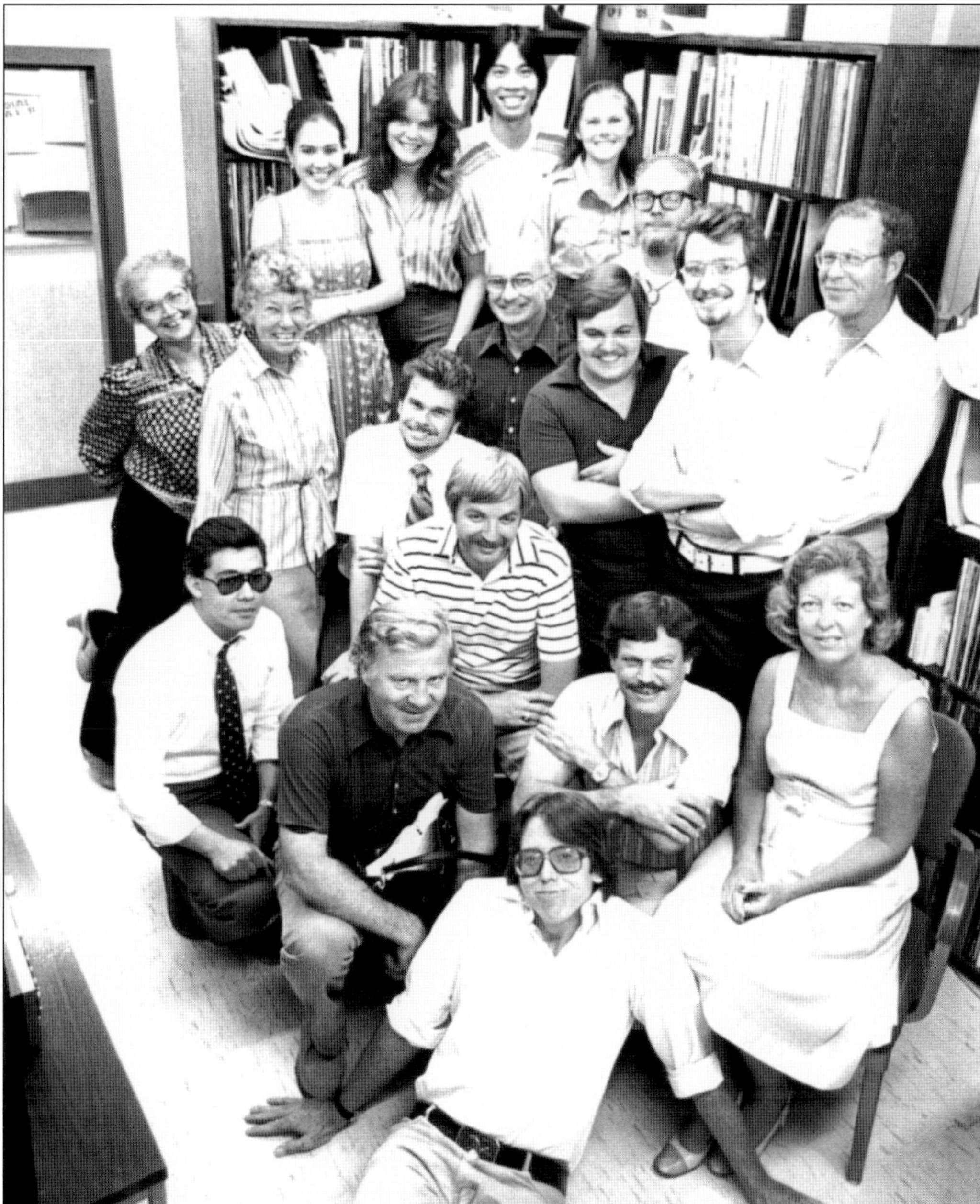

This 1970s photograph features KUAT staff members. Classical host Bill Pitts is seated on the floor in front. Pat McQuown (in the third row, standing, wearing striped shirt) was the program manager, and Ed Kupperstein (third row center in the dark shirt) was the station manager. James Reel (standing second from right in the third row wearing a white belt) was a classical music host and music director. It takes a talented group of individuals to efficiently run a radio station, whether it is listener-supported or commercially owned. In the background, a portion of KUAT's large classical music record library can be seen. (John Kelley, Arizona Pubic Media.)

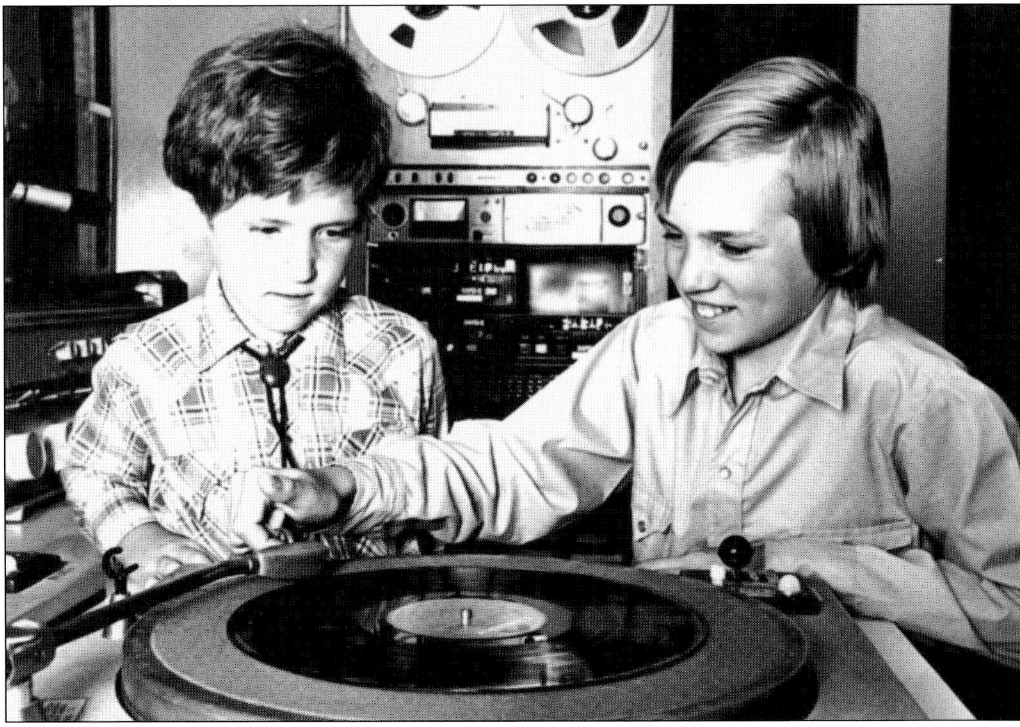

Gathering listener participation for NPR and local programs was critical to a station's success, as were the promotional contests and fundraising campaigns. Larry Schnebly, a veteran Tucson broadcaster, joined the Arizona Public Media Advisory Board to help raise money to support the Tucson public radio and television stations. In this photograph, taken inside the KUAT radio control room, are Tucsonans Jeffrey Hufford (left) and older brother Chris, both of whom were winners in KUAT-NPR's Young People's Radio Festival. (John Kelley, Arizona Public Media.)

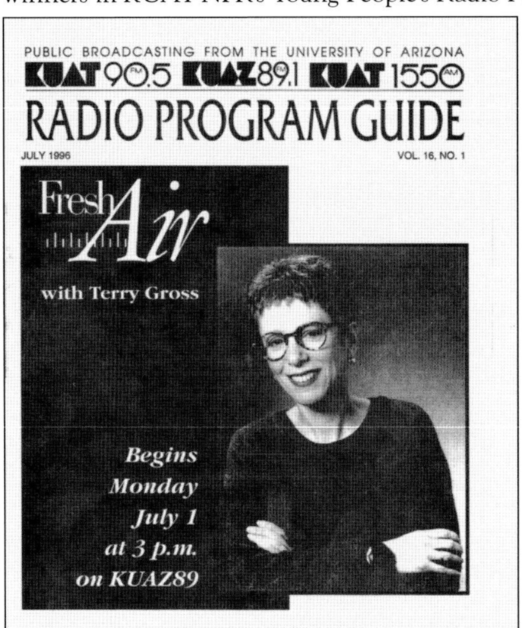

NPR started with the *All Things Considered* afternoon program of news reports and commentary, which continues to be heard on KUAZ AM-FM every afternoon. In the morning hours, it is *Morning Edition*, which also has a weekend program. Local news and weather inserts are aired as well. NPR has expanded over the years and provides local stations with national music entertainment, news, and discussion programs. This advertisement, from the July 1996 KUAT Radio Program Guide, shows Terry Gross's *Fresh Air* interview program, which was added to the KUAZ radio schedule. (John Kelley, Arizona Public Media.)

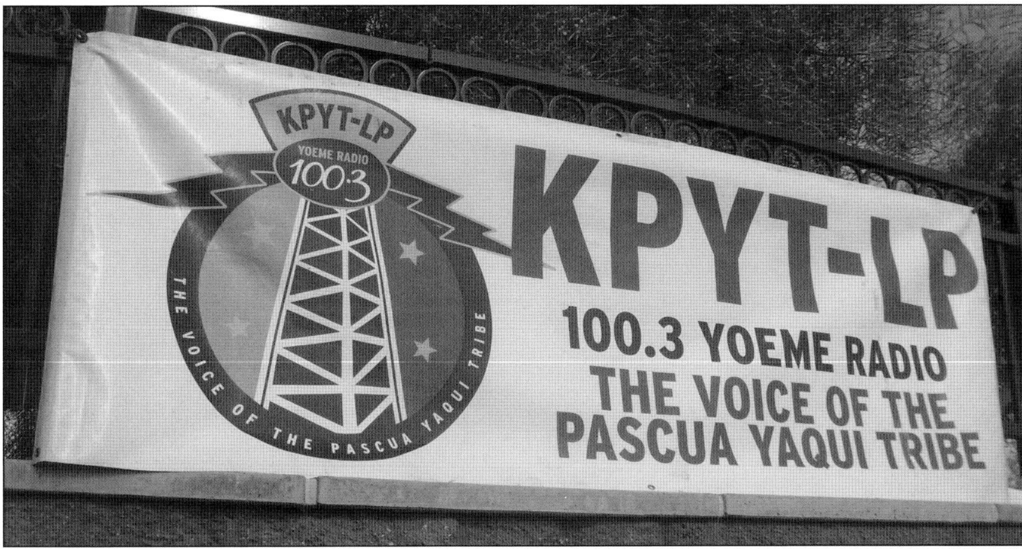

New public radio stations emerged that were not NPR network affiliates. In 1983, KXCI came on at 91.3 with a "progressive," noncommercial format. Housed in an old brick house on South Fourth Avenue, KXCI offers a diverse music offering and features the daily *Democracy Now!* program of news and commentary hosted by Amy Goodman. When the FCC made it possible for low-power community radio stations to come on the air, the Pascua Yaqui Tribe started KPYT-LP on 100.3 FM, with 100 watts, in 2005. (Hector Youtsey.)

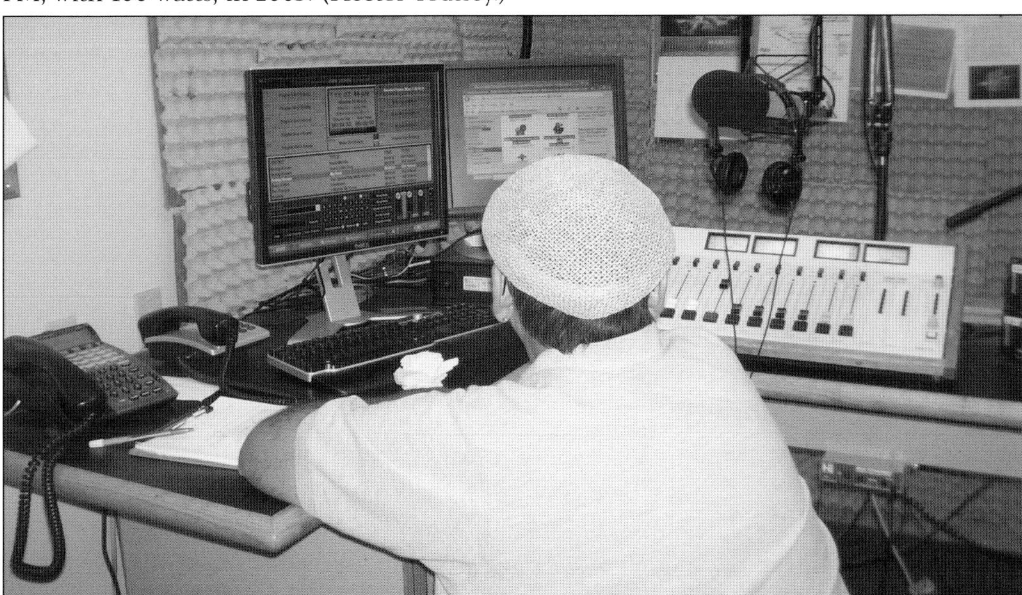

This is the control room at KPYT, where 30 volunteers produce public-service announcements and host programs such as *The Diabetes Prevention Program*, *Club House Radio*, *The Education Hour*, *Vahi Vo'om Hiawai* (entirely in the Hiaki language), *The Sewa Uusim Radio Show*, and *Tribal Legacies*. Music programs include *The More Variety Morning Show*, *The Backyard Jam*, *Country Jubilee*, *En Concierto*, *The Snake Break*, and *The Classic Rock Block*. The music genres heard on the station include Tejano, Yaqui music (folk and Paskola), old skool, plus oldies and classic rock, all designed specifically for its audience. (Hector Youtsey.)

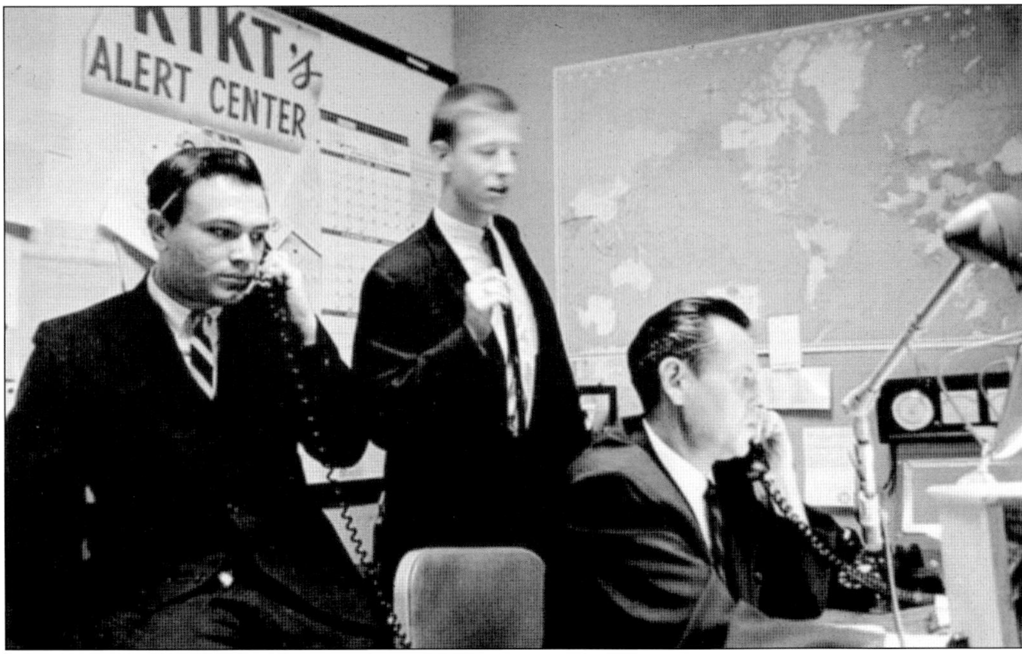

When KTKT took on the Color Radio Top 40 format in 1957, newscasts were heard at the top of each hour. This more polished and formatted model was developed originally by Gordon McLendon at KLIF in Dallas. Pictured here is the 1963 KTKT news Alert Center, with, from left to right, Gene Adelstein, Joe Bailey, and news director Lloyd Couch. (Ray Lindstrom.)

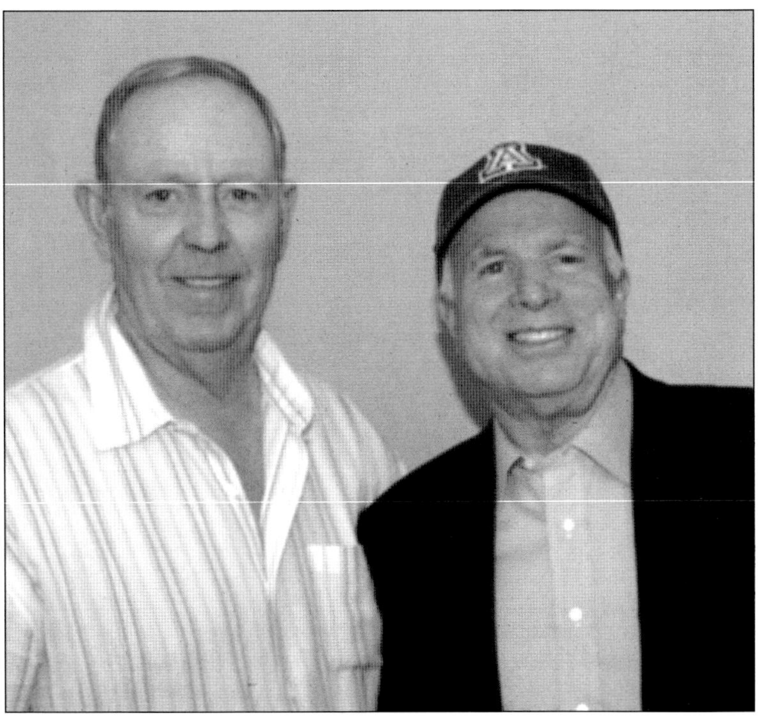

Government officials and campaigning politicians are regularly asked questions in interviews on radio newscasts and talk programs. KCEE radio newscaster Clare Coffman frequently interviewed U.S. senator Barry Goldwater. In this photograph, Sen. John McCain (right) has been interviewed by KCUB's Jim Slone. (Jim Slone.)

KTKT NEWS "CHANNEL 99"

Coverage of the Kennedy assassination by KTKT News is remembered by Ray Lindstrom: "President John F. Kennedy was assassinated on November 22, 1963 in Dallas, Texas. It was about 11:30 in the morning and I was driving to my political science class at the University of Arizona. During the day, I was getting my college degree. At night I was a disc jockey at KTKT radio. I was listening to my station, KTKT. There was a Kinney Shoe commercial on the air. All of a sudden Lloyd Couch, the news director, interrupted the commercial with a news bulletin. You never interrupted a commercial! Any bulletin could wait a few seconds to make sure you got those few extra bucks in the coffers. But, not this news bulletin. . . . I decided to go to the radio station, a magnet for any radio guy at a time like this. When I got there, the three executives were in the production studio having a meeting. Phil Richardson, station manager, Frank Kalil, assistant manager, and Jerry Stowe, program director. The president of the United States had just been shot and they were deciding how to format the station appropriately until at least the state funeral was over. The question was, 'How can we go ahead and play rock and roll music during this time of national tragedy and mourning?' The answer was, they could not. They all agreed to stop the Top 40 format immediately and start playing patriotic tunes until after the funeral." (Ray Lindstrom.)

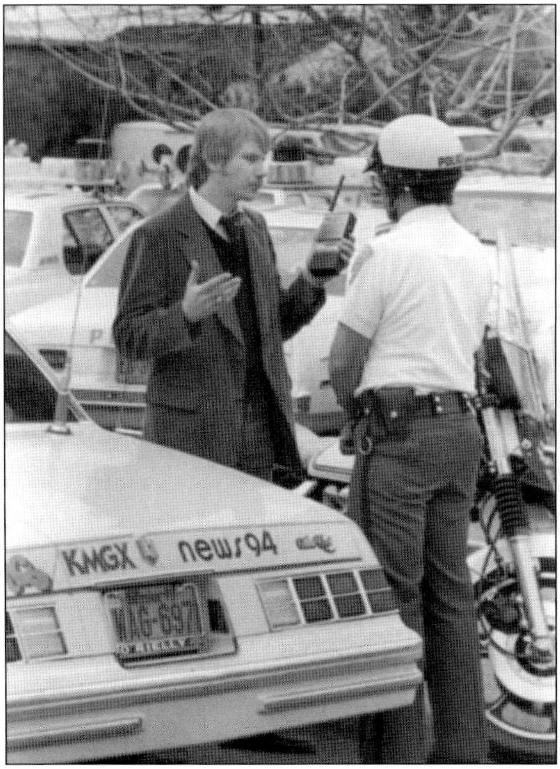

Radio station mobile news units came in different shapes and sizes. Bob Lee, a former television newsman, became a disc jockey, covering the news at KMGX 940. Here, he interviews a Tucson police officer at the scene of a story. Most of the mobile units were vehicles the stations acquired through a "trade-out," meaning the car dealer swapped the vehicle for advertising in lieu of the station paying cash. In the late 1950s, KTKT's Mobile News Patrol was a yellow van equipped with red lights so it could go "Code 3." Because of that capability, the Pima County sheriff's office could ask EMT-trained reporter Glen Hatfield to treat the van as an ambulance, and his reports made for exciting radio. (Bob Lee.)

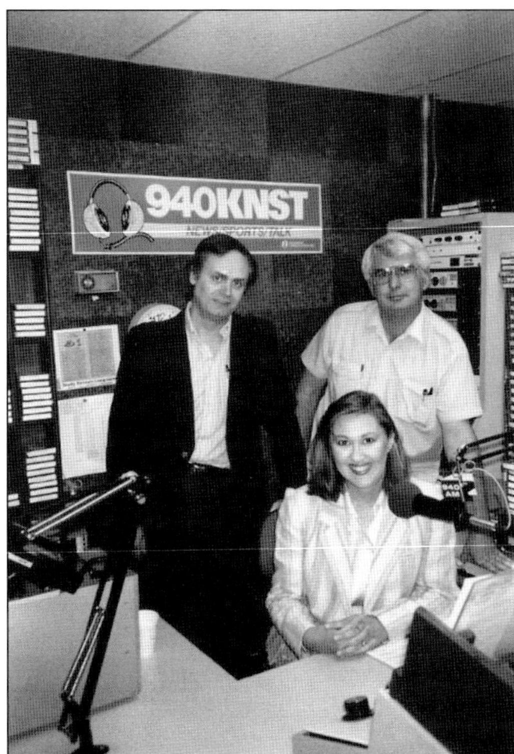

KMGX, later called KNST, was the first station to provide continuous news blocks during morning and afternoon drive times. News director Bob Lee, Tonie Stanton, and Brian Jeffries did the morning program. Bob Ritter, (right, with glasses) was the afternoon host. The interviewees shown here are unidentified guests. The 1982 KNST studios and newsroom were in Placita Village downtown. (R Dub!.)

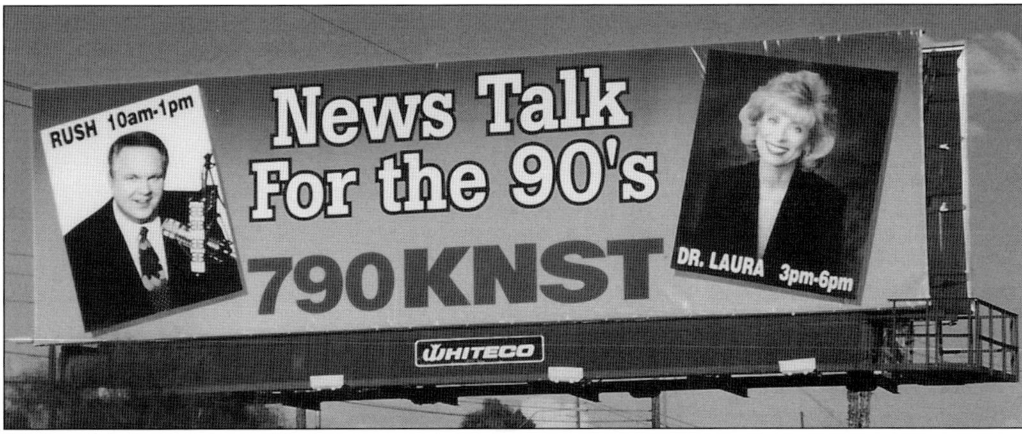

Talk radio has been called the savior of AM radio stations, as Larry King and Rush Limbaugh proved that programs discussing national issues and politics could draw big audiences. King was on overnights, and Limbaugh came along in the late 1980s with his three-hour daytime program. Within a few years, Limbaugh was the highest-rated talk-show host in the country, heard on 600 radio stations. Around the same time, listeners were introduced to Dr. Laura Schlessinger, who confronted personal and social issues during her three-hour daytime program, heard on 450 stations. KNST carried both programs and put up billboards like this one to promote them. (R Dub!.)

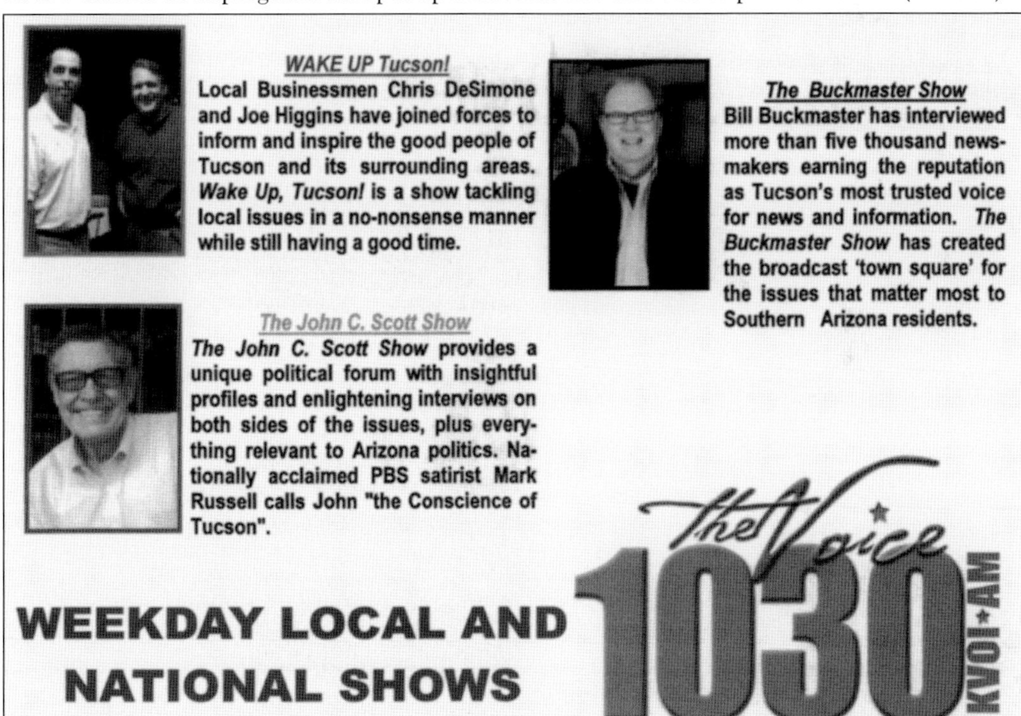

The growth of talk radio brought forth more national personalities, most of them political conservatives, from G. Gordon Liddy to Michael Reagan. There was room for some local talk shows, mostly in the morning and afternoon news blocks. In Tucson, the three hosts pictured in this KVOI advertisement had long-running programs discussing community issues and interviewing local personalities. Talk on FM was exclusive to public radio, but some commercial stations tried it, like FM 104.1 in Tucson, with its morning host, Jon Justice. (Doug Martin, KVOI.)

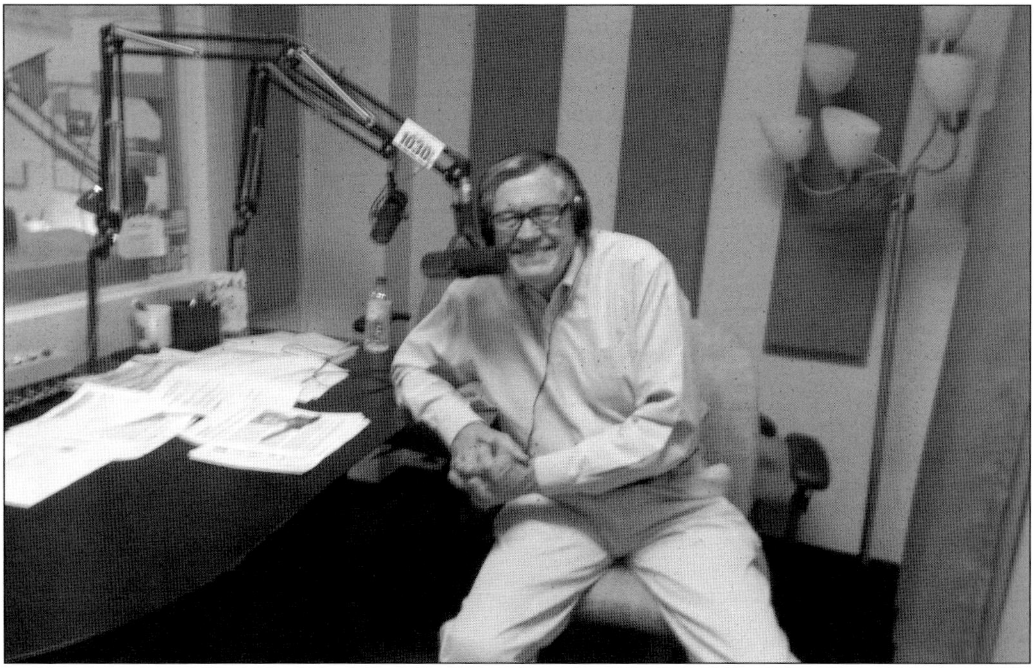

John C. Scott has been a news-talk personality on Tucson radio for decades. He is one of the KCUB Bunch pictured on the inside cover of this book. His talk-radio career started at KTUC in 1989, then continued at KTKT. His program is a family affair, with his son Mark and wife, Amy, working to produce it. He was on the air on KJLL at the time of the attempted assassination of Congresswoman Gabriel Giffords in January 2011, and C-SPAN-TV chose Scott's program to broadcast nationally. Here, he is conducting his two-hour afternoon program on KVOI 1030. Scott is a winner of the Tucson AdFed Golden Mic award. (RJ.)

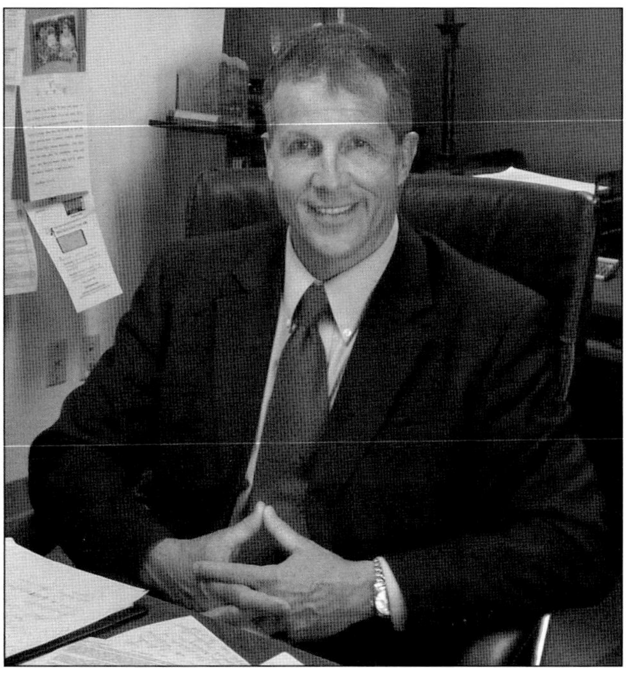

Good News Broadcasting, a Christian faith-based company, became the last group of radio stations that were locally owned. Doug Martin (shown here) is president and general manager of the four stations, each having a different format. Religion's presence on radio goes back to the early days, when programs like NBC's *The Eternal Light* and ABC's *Old Fashioned Revival Hour* were heard nationwide. In Tucson, the Trinity Presbyterian Church was on KCNA, as was the First Assembly of God, which had Sunday and daily programs. (RJ.)

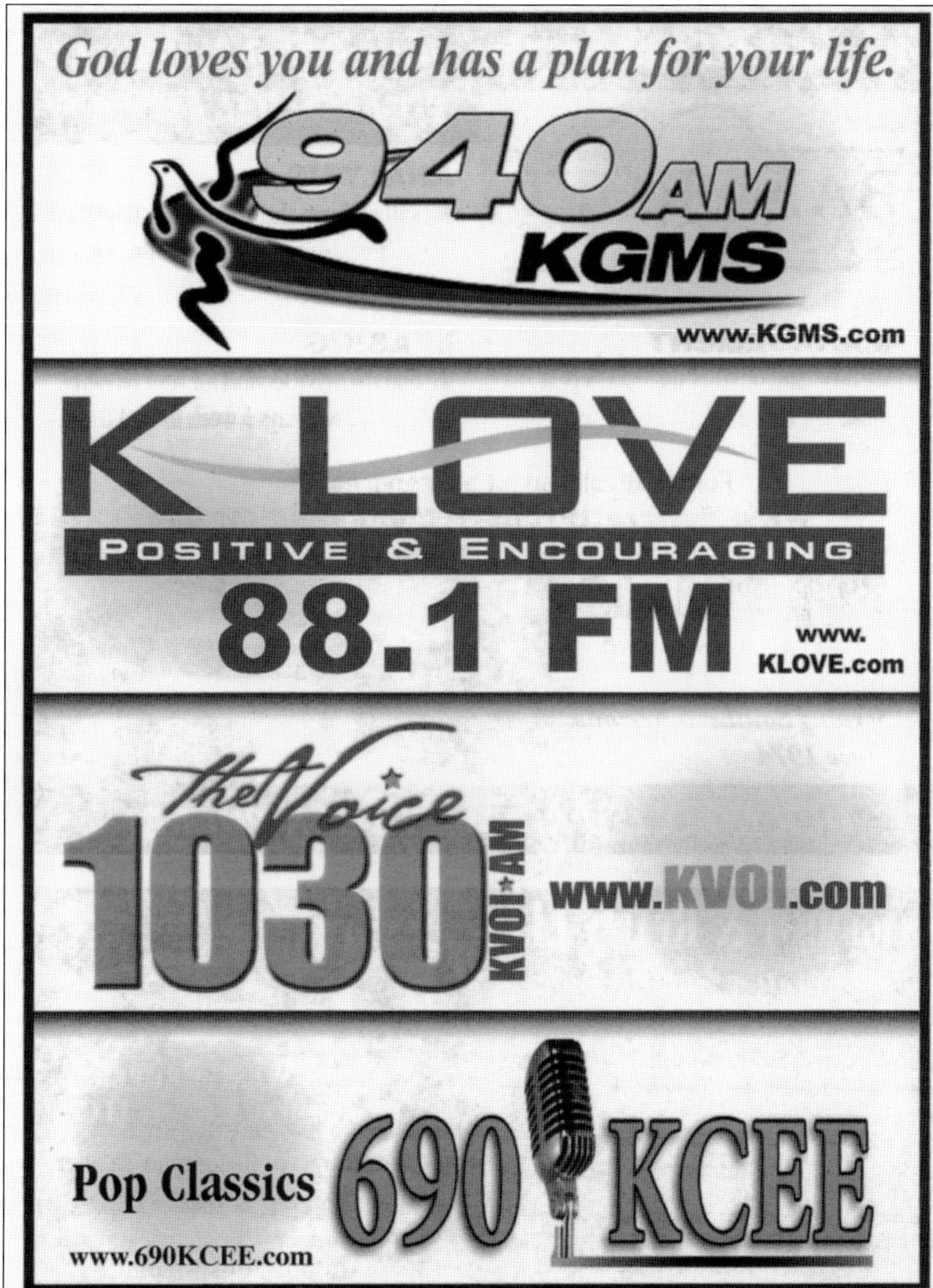

KGMS 940, once country music's KHOS and the original KNST, is the outlet for the traditional religious preaching programs and their loyal listeners. K-LOVE 88.1 on FM follows a syndicated contemporary Christian music format. KVOI on 1030 AM is a news-talk station that carries three daily local talk programs, plus national conservative hosts like Mike Huckabee and Dennis Miller. The KCEE call letters have been around since 1957, and this station, at 690, now carries nostalgia music, following its acquisition from Jim Slone. (Doug Martin.)

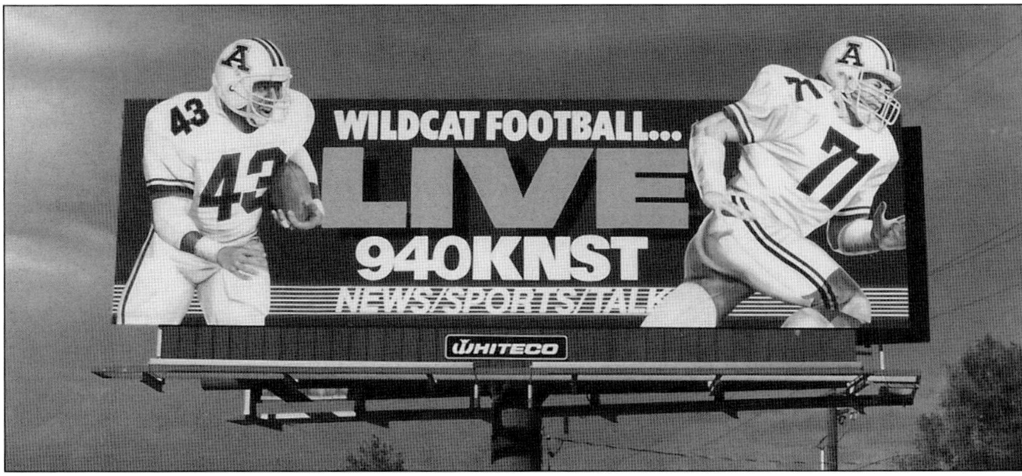

University of Arizona sports events have always been covered on Tucson radio. Prior to television, football games could sometimes be heard on three radio stations simultaneously. Since the 1980s, they have been heard on KNST, as seen on this billboard, then moved to KCUB, with its all-sports format. Sportscasters travel with the teams, and radio play-by-play is an important part of reaching fans. Game broadcasts are heard all over the state, with alumni able to follow the games on stations in Phoenix and other cities. Tucson has had two all-sports radio stations, ESPN affiliate KFFN 1490 and KCUB 1290. (R.Dub!.)

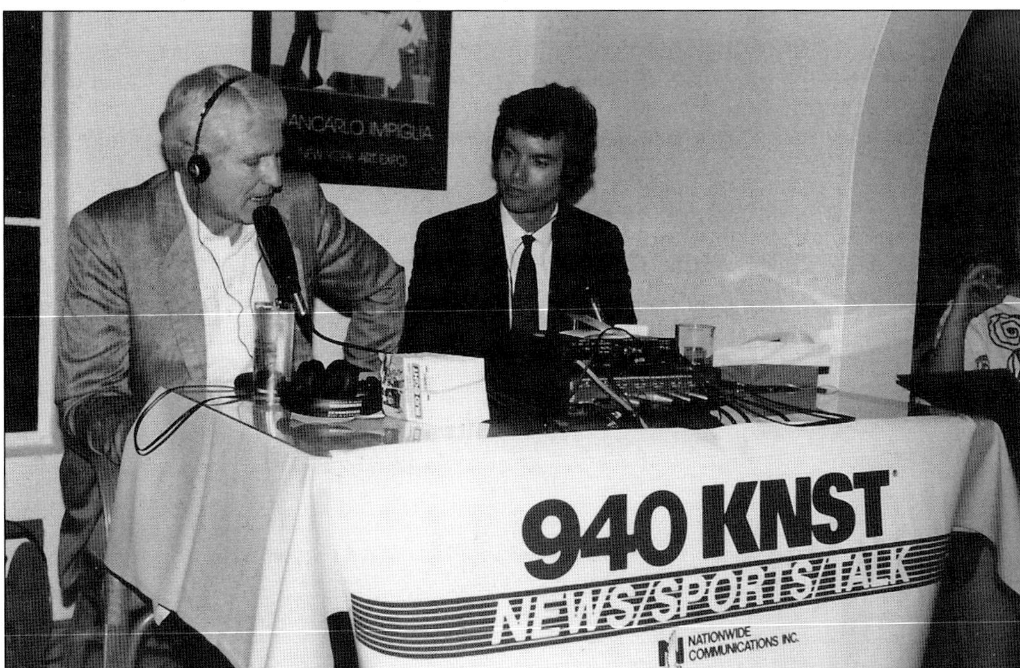

For over two decades, the radio "Voice of the Wildcats" has been Brian Jeffries. He started as a disc jockey, then moved into sports reporting and talk shows. He has interviewed coaches and players from all of the UA sports teams. The university has had championships in many sports, but none were as big as the 1997 basketball national championship team, which was coached by the legendary Lute Olson. In this photograph, Jeffries (right) is interviewing Coach Olson "back in the early days" for both of them. (R.Dub!.)

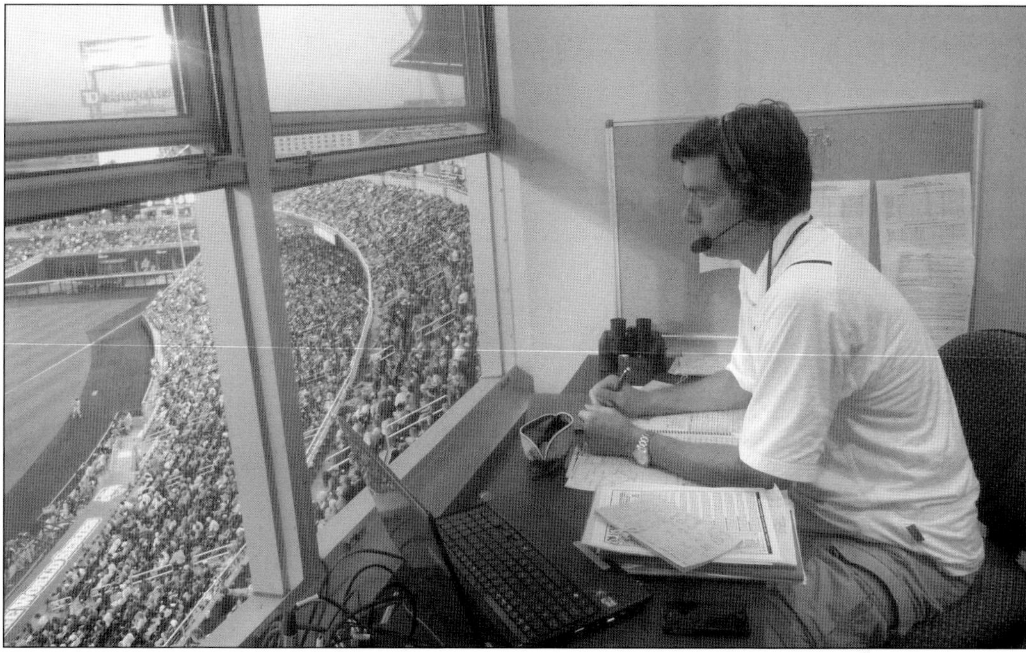

Calling play-by-play for a football game on radio means letting the listeners "see" the game, and Brian Jeffries has been doing that from high above the field at the UA stadium. Jeffries was inducted into the Arizona Broadcasters Hall of Fame in 2012. (Brian Jeffries.)

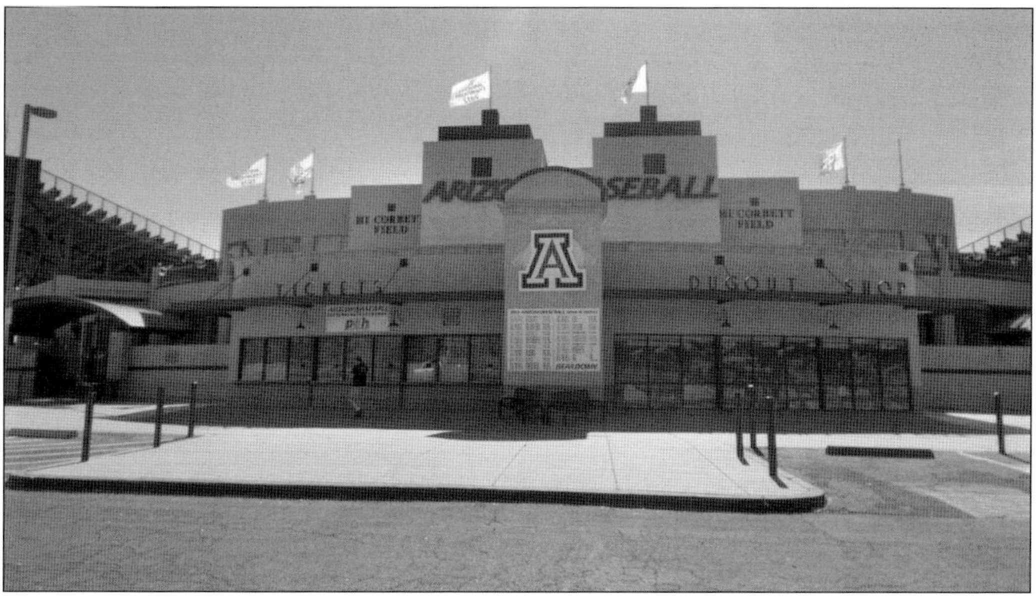

Hi Corbett Field has been the home for local baseball games since the 1940s. The minor-league Tucson Cowboys played there, and Tucson radio sportscasters covered all the activity. Major-league teams arrived to train in Tucson each March from 1948 until 2009. While most baseball games moved to the new Tucson Electric Park, Corbett Field continues to be utilized and remains the home park for Wildcat baseball. The versatile Brian Jeffries has broadcast UA baseball games from there on radio. The team has competed in the College World Series, winning the national championship in 2013. (RJ.)

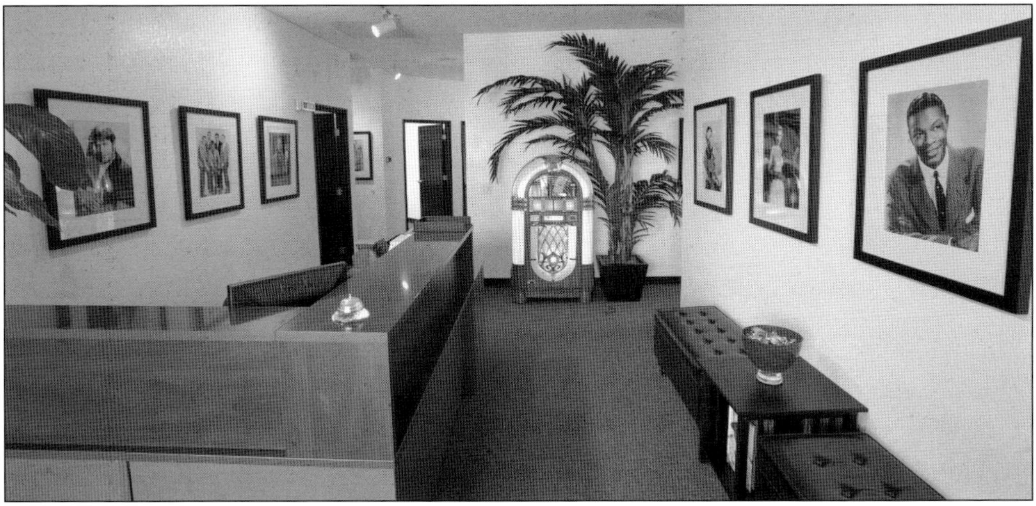

Jim Slone was not out of broadcasting for long. He sold his station group in 2001, but in 2006, he purchased KEVT 1030. He resurrected the KCEE call letters, along with the nostalgia format. The above photograph shows the station's lobby, which features photographs of artists like Nat "King" Cole (right), whose single "That Sunday, That Summer" was a huge hit in 1963. In the center of the KCEE lobby is Slone's colorful 1946 Wurlitzer jukebox, which plays only 78 rpm records! The playlist below is indicative of the wide variety of songs that Slone selected. Although he sold the station in 2009, the new owner, Good News Broadcasting, has continued his format, moving it to 690 on the dial. (Above, Jim Slone; below, Doug Martin.)

JUKEBOX JAMS

Frank Sinatra *My Way*

Perry Como *As Time Goes By*

Johnny Mathis *Chances Are*

Johnny Cash *I Walk the Line*

Mamas and the Papas *California Dreamin'*

Linda Ronstadt *Different Drum*

Neil Diamond *Sweet Caroline*

Billy Joel *She's Always a Woman to Me*

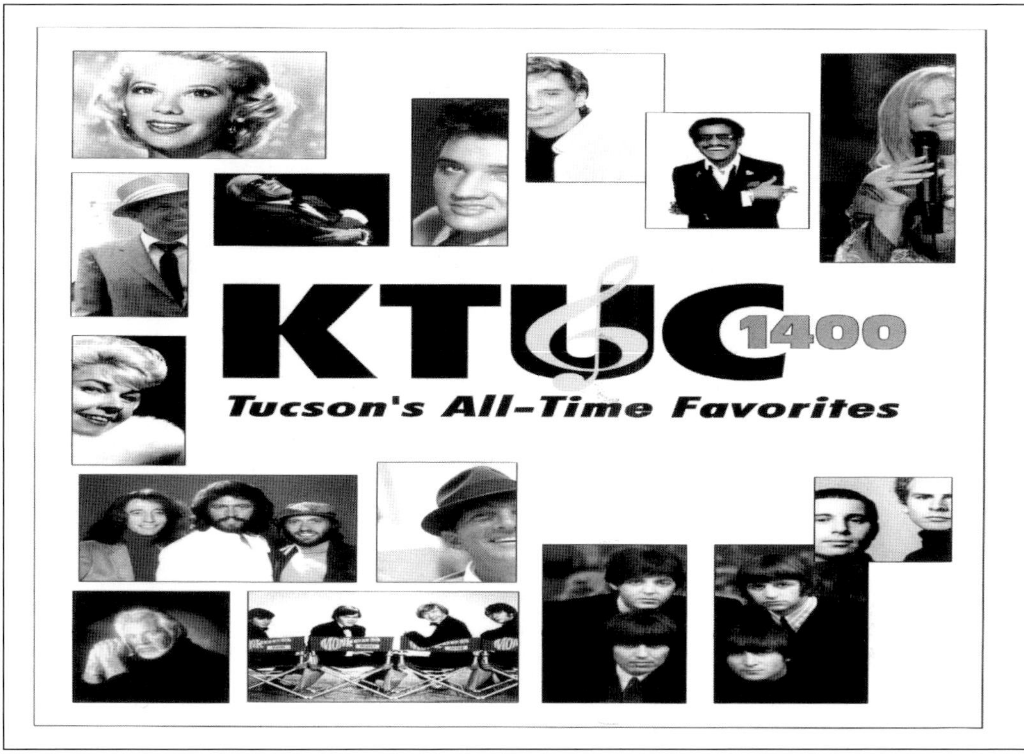

The nostalgia format has been very popular in Tucson. In addition to KCEE, stations KGVY, KSAZ, and the historic KTUC have programmed songs from the 1940s through the 1970s. Joe Crystal, the longtime KOLD manager, put KGVY on 1080 in 1981. Many veteran broadcasters have worked there, including Dan Gates and Bill Mortimer. (Cumulus Broadcasting.)

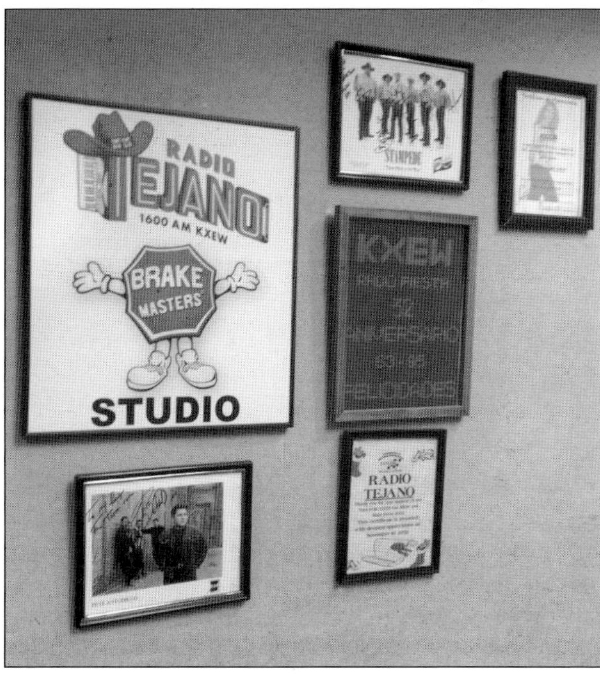

Spanish-language broadcasting in Tucson started with Jacinto Orozco on KVOA, and it grew with its first full-time station, KEVT. Since then, several stations have appealed to that audience. The second station was daytime-only KXEW 1600, which came on the air in 1963 and was managed by Ernesto Portillo. The KXEW call letters continue, now owned by Clear Channel Broadcasting. The station has a Tejano music format, as reflected in this photograph of pictures hanging in the station. Other Spanish stations came along, including 1210, which took the KEVT call letters, 102.1 KCMT on FM, and even the 580 and 990 frequencies. Lotus Broadcasting owns KCMT and took KTKT to Spanish as an all-sports station, ESPN Desportes. (RJ.)

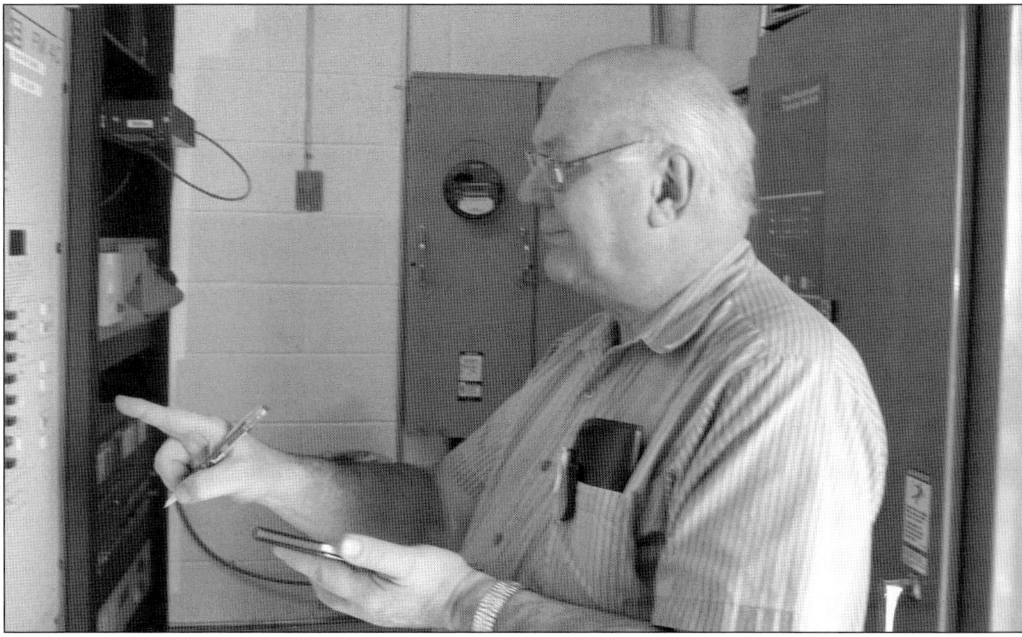

The role of the broadcast engineer is of equal importance to any other member of the radio station's staff. Broadcast equipment must not only be installed and operated, but it must also receive constant maintenance to keep the station on the air uninterrupted. Barry Mishkind is one of the professionals who do that work. In this photograph, he is working with the transmitter for station KOHN, a unique radio station that he helped build. It is a 10-kilowatt "public" FM station at 91.9, located near Sells, west of Tucson. It programs for the large Tohono O'odham nation. (Barry Mishkind.)

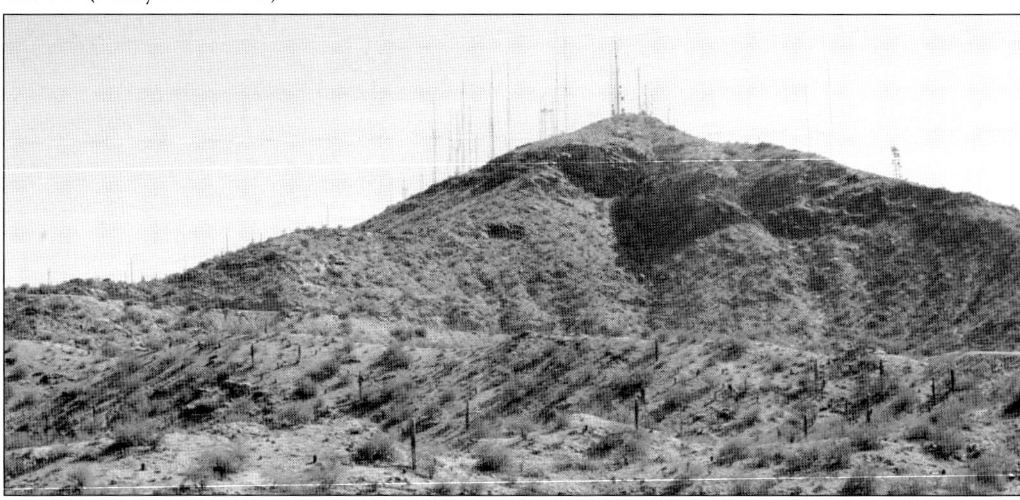

Barry Mishkind helped build the Tucson Mountain site (shown here), which consolidated in one place most of the FM and some television transmitters and antennas and other communications systems. In addition to his engineering duties for the stations that contract with him, Mishkind has written many articles for *Radio World* magazine and developed Internet websites. These activities have earned him the title "the eclectic engineer." His website, www.oldradio.com, contains information of interest to fellow engineers and has a large section devoted to the history of radio stations in many US cities. (RJ.)

ARIZONA BROADCASTERS ASSOCIATION HALL OF FAME

ABA
Arizona Broadcasters Association

HALL OF FAME INDUCTEES

1990
Rex Allen*
Tom Chauncey*
Dick Gilbert*
John R. "Jack" Williams*

1991
Virginia Mittendorf*
Willard Shoecraft*
"Wallace & Ladmo Show"
(Lad Kwiatkowski*, Pat McMahon & Bill Thompson)

1992
Lew King*
Homer Lane*
George Wallace*

1993
Bill Lester*
Jack Murphy*
Buck Owens*

1994
Steve Allen*
Jewell Lewis*
Ray Smucker*

1995
Chuck Blore
Jeanne Metzger Michael*
John Redfield*

1996
Joe Crystall*
Joe Dana
Bea Hoeksama*
Ira Lavin*

1997
Ken Heady*
Delbert Lewis
Mary Jo West

1998
Gene Autry*
Guy Christian*
Bill Close
Jack Jacobson*

1999
Bob Ellis
Larry Schnebly
Lee Shoblom
Ray Thompson*

2000
Dick De Angelis*
Gary Edens
Frank Kalil
Joe Tabback*

2001
Rita Davenport
Ray Odom
Jim Slone

2002
C.E. "Pep" Cooney*
Phyllis Ehlinger*
Lou Silverstein

2003
Karl Eller
Larry John
Jack Parris

2004
Jack Clifford
Bill Heywood*
Al McCoy
Jon Ruby*

2005
Pat McMahon
Bill Miller
Roy Track*
Charlie Van Dyke

2006
Bob Allingham
Hugh Downs
Norm Ginsburg*
Marge Injasoulian

2007
Chuck Artigue
Bob Lee
Beth McDonald

2008
Jim Arnold
Ron Bergamo*
Betty Waters*
Preston Westmoreland
Gary Steven Woodworth*

2009
Kent Dana
Bud Foster
W. Steven Martin
Dave Pratt

2010
Guy Atchley
Ned Foster
Joe Garagiola Sr.

2011
Bill Austin*
Patti Kirkpatrick
Tim Hattrick
Willy D. Loon

2012
Tom Chauncey II
Brian Jefferies
Linda Turley

The Arizona Broadcasters Hall of Fame was created to honor the many outstanding people who have dedicated their careers to the radio and television broadcasting business in the state. This list shows all of the inductees, including many Tucsonans. How is a radio station deemed to be "successful?" Within the business, success is based on high ratings, reflecting the number of listeners, how long they listened, and their ages. This information is important to the stations and the advertising agencies and is the basis of commercial sales. Sponsors seek to put their advertisements on the stations that most appeal to those who might spend money on their products or services. (ABA.)

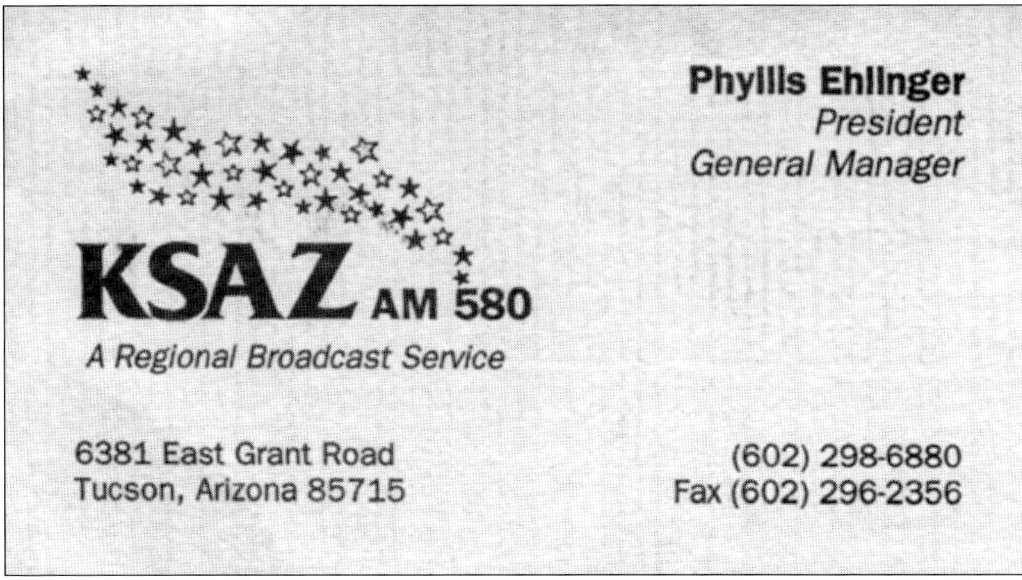

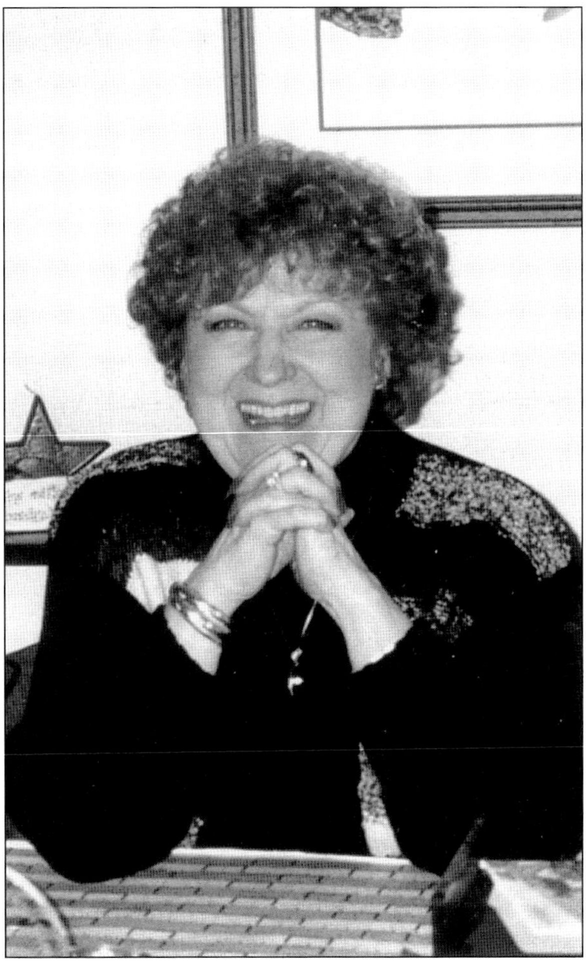

KSAZ 580 was the station that evolved from KCNA-KTAN and KIKX. When the frequency was opened to new applicants by the FCC in 1994, the winner was Phyllis Ehlinger. She became the first woman to own a radio station in Tucson. To get the station on the air, she had to have the transmitter relocated. Its three towers were constructed just northwest of Marana, which became the official "city of license." Ehlinger chose the call letters KSAZ. The program format put KSAZ in competition with the other nostalgia music stations. KSAZ signed up with the ABC satellite network and its nationally known personalities. Ehlinger had been in radio for many years, serving as both a successful advertising salesperson and general manager following her move to Tucson in 1960. She became known as the "queen of radio broadcasting" by her peers. She was inducted into the Arizona Broadcasters Hall of Fame in 2002, and upon her passing in 2006, the Tucson Ad Club announced that an annual award would be given in her honor to a deserving student. KSAZ was eventually sold, becoming the latest convert to Spanish-language programming. (Above, RJ; left, Advertising Federation.)

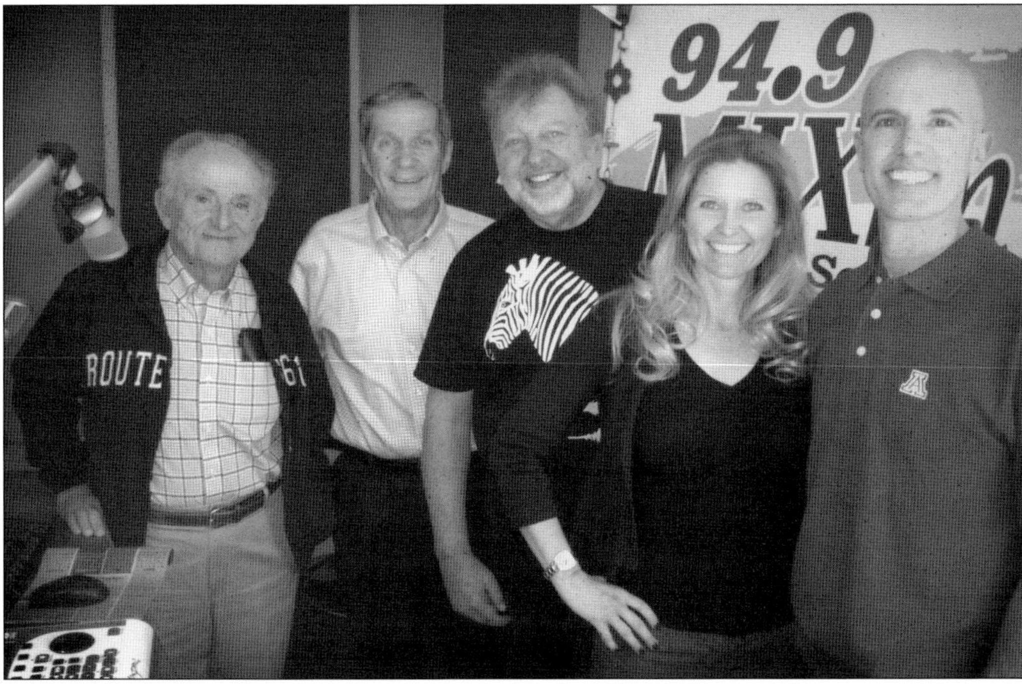

Tucson has had many outstanding morning radio personalities who have remained on the air for years. KMXZ 94.9, the "soft rock" station, has had Bobby Rich and his "Morning Mix" crew of fun and games for over two decades. In this March 2013 photograph, Rich is in the center with his trademark big smile. In 1993, Rich came to Tucson from a successful "morning zoo" show in San Diego. Posing with Rich are, from left to right, the visiting co-author of this book, Russ Jackson; Journal Broadcast Group general manager and legendary Tucson air personality "Sunny Jim" Arnold; and Rich's program cohosts, "Mrs. Grant" and Greg Curtis. Arnold was inducted into the Broadcasters Hall of Fame in 2008, and Rich was inducted in 2013. (Marty Bishop, KMXZ.)

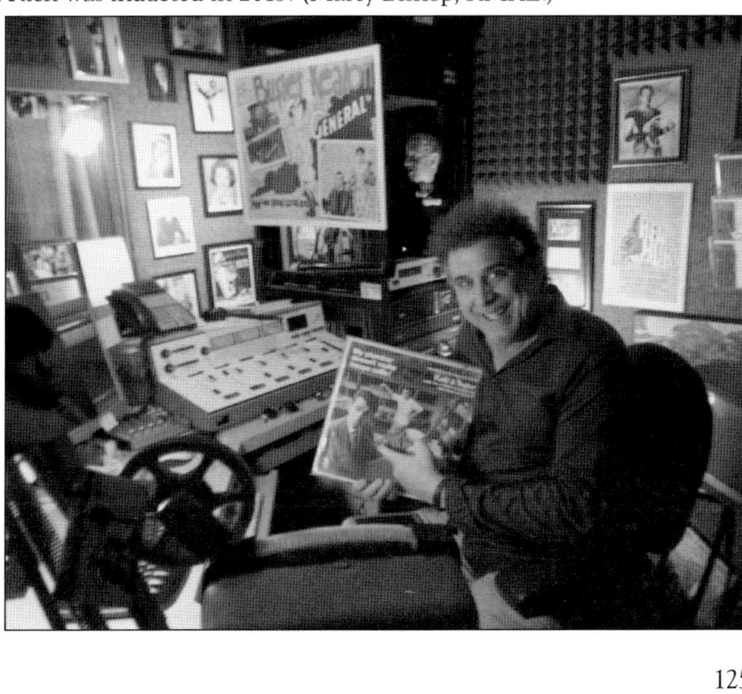

KVOI production director Gerry Cross pays tribute to Frank Kalil. Seen here in his studio, Cross holds one of his favorite albums, a 1966 live comedy album that Kalil recorded with future adman Jay Taylor, titled, *My Plumber Doesn't Make House Calls*. Kalil's radio shows were known for his one-liners. He was inducted into the Broadcasters Hall of Fame in 2000. (RJ.)

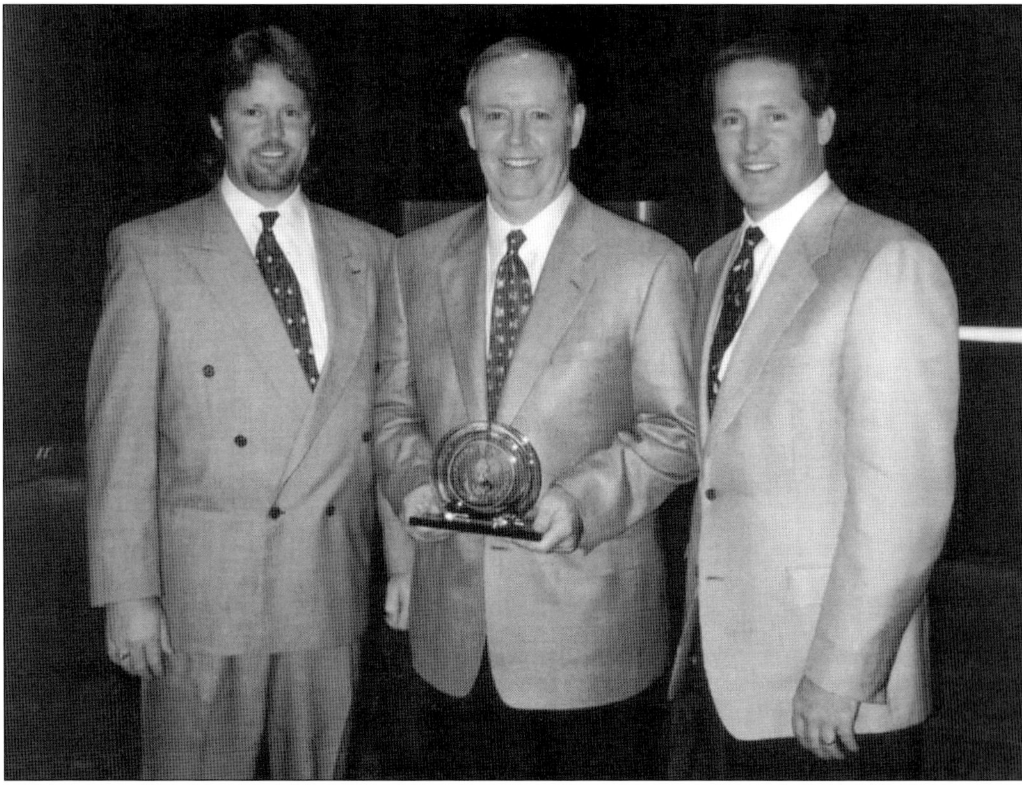

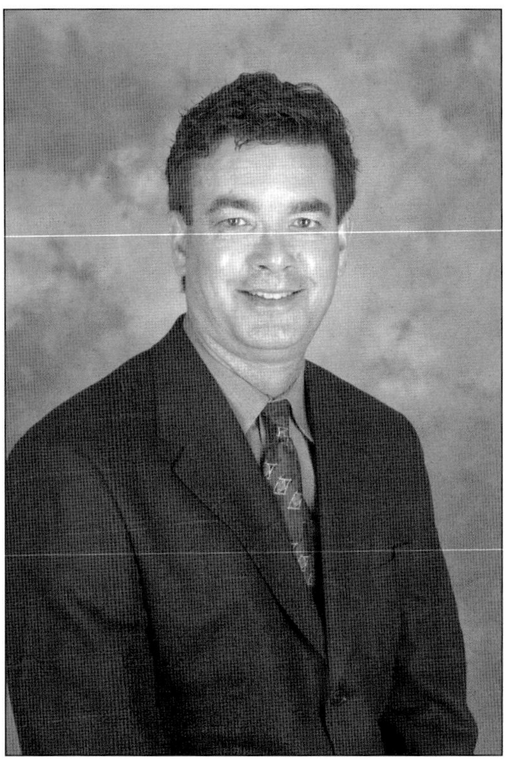

Retired station owner Jim Slone was inducted into the Arizona Broadcasters Hall of Fame in 2001. In the above photograph, he shares his award with sons Fred (left) and Jamie, following the award luncheon in Phoenix. At left is "The Voice of the Wildcats," Brian Jeffries, who was inducted in 2012. All of the hall of famers were recognized for their career contributions to the industry. Those inductees with Tucson connections have been spotlighted at various places in this book, and the full list appears on page 123. (Above, Jim Slone; left, Brian Jeffries.)

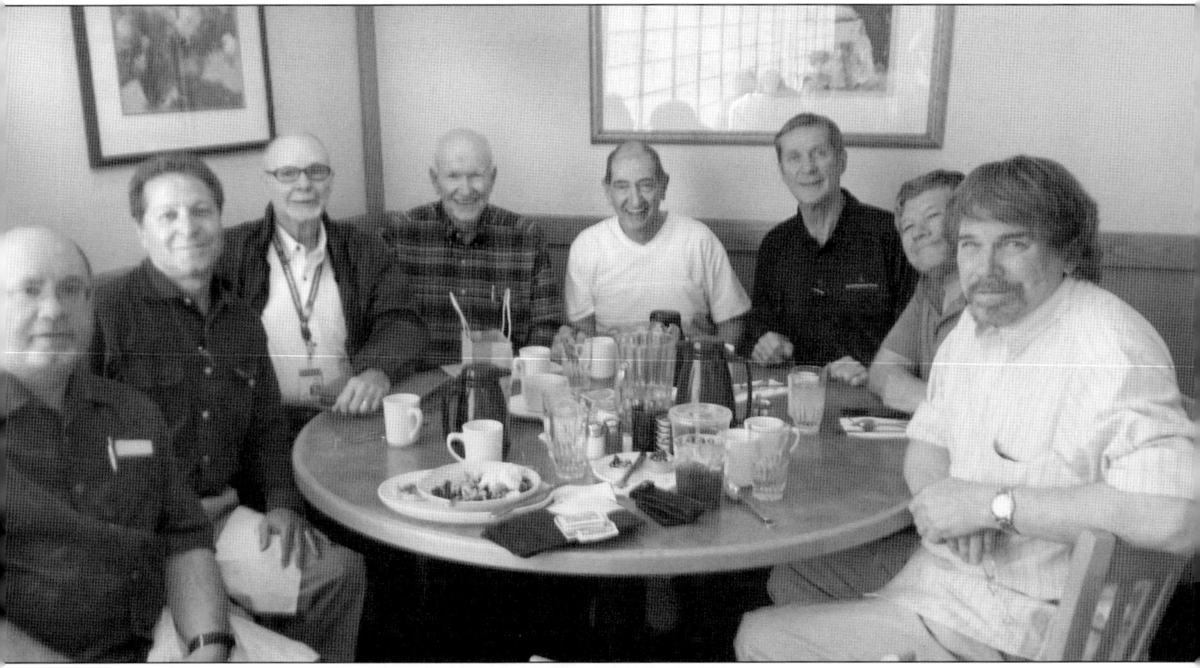

In 1963, a group of executives from KGUN-TV started getting together for weekly coffee at a local restaurant to exchange information and, as Larry Schnebly, one of the founding members, says, "tell a lot of bad jokes." The group calls itself the "Coffee Crew," and it has been meeting Mondays and Fridays without interruption ever since. The group consists of many retired broadcasting and advertising professionals, as well as those currently employed. This photograph shows those in attendance for coffee on March 15, 2013. They are, from left to right, Dave Slavin, Rich Moret, Jack Parris, Ray Lindstrom, Larry Schnebly, "Sunny Jim" Arnold, Tommy Gallego, and Bob Lee. Past Coffee Crew members have included Frank Kalil, George Wallace, Jack Jacobson, Jake Crellin, Art Waller, Fred Vance, and Chris Borden. In a 2009 *Tucson Weekly* article, writer Lee Allen said, "Each of the members brings roughly a half-century of on-air experience. They usually flip a coin to see who pays for everyone's coffee, a pattern George Wallace started 40 years ago." (RJ.)

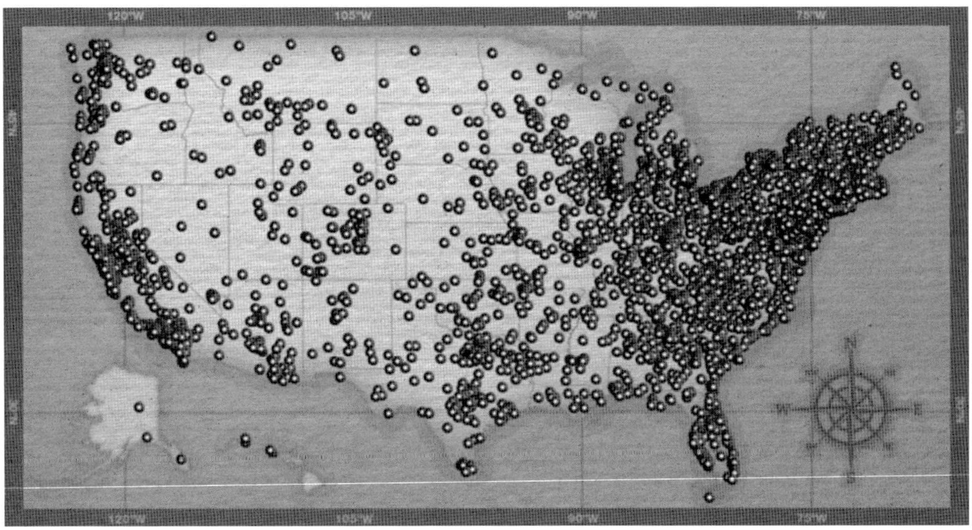